Bill Statham

WHAT'S *REALLY* IN YOUR BASKET?

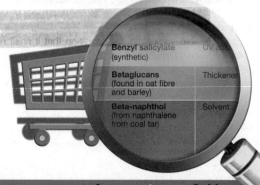

Benzyl salicylate (synthetic)	UV abs
Betaglucans (found in oat fibre and barley)	Thickener
Beta-naphthol (from naphthalene from coal tar)	Solvent

An easy-to-use Guide to Food Additives and Cosmetic Ingredients

summersdale

WHAT'S REALLY IN YOUR BASKET?
Previously published as THE CHEMICAL MAZE by
Summersdale Publishers Ltd, 2006
First published by POSSIBILITY.COM in April 2001

Summersdale Publishers Ltd
46 West Street
Chichester
West Sussex
PO19 1RP
UK

www.summersdale.com

Printed and bound in Italy.

ISBN: 1-84024-607-3
ISBN 13: 978-1-84024-607-0

Contents

About the Author

 Bill Statham lives with his wife and business partner Kay Lancashire in Victoria, Australia. He is a researcher, writer and publisher with an interest in health education and is committed to making a positive difference to the health of people and the environment.

He studied and practised homeopathy both in Australia and England for over ten years. During this time he became increasingly concerned about the detrimental effects on people's health caused by synthetic chemicals in the foods we eat and products we use every day.

Bill wrote *What's Really in Your Basket?* to make it simpler and easier for people to recognise those additives and ingredients in foods, personal care products and cosmetics having the potential to cause discomfort and ill health. With this recognition comes freedom of choice, and for many a new lease of life.

Foreword

Both new and experienced 'safe-consumers' will benefit from Bill Statham's research and guidance in *What's Really in Your Basket?*.

Do not be deceived by its miniature size… this little handbook could create BIG changes in the way you look at food and personal care forever. If you take Bill's advice to heart, no longer will you be able to participate in mindless shopping-trolley expeditions… ignorance may once have been bliss; but now it spells danger in our mass-manufactured, profit-driven, long shelf-life, chemical-romanced society.

After spending many years researching the toxic ingredients in skin and personal care, and successfully avoiding all of them in the products I create, I am happy to recommend Bill Statham's *What's Really in Your Basket?* as an excellent reference guide to anyone questioning the safety of those strange-sounding 'naturally derived' ingredients in their skin and personal care.

Moreover, as a long-term Certified Organic consumer, I truly hope you put your money where your health is, and 'Go Organic'!

Narelle Chenery
Director of Research and Development, Miessence

How to Use Your Guide

The reference part of this book is divided into two sections.

Section one provides an alphabetical list of **food additives** approved in the EU together with their E numbers. If you know only the E number of a food additive, then refer to the E Number Conversion Chart to find its name.

Section two lists in alphabetical order ingredients that may be found in **cosmetics and personal care products**. Some cosmetic and personal care ingredients are also approved as food additives, and therefore may be found in section one. Examples are lactic acid (E270), citric acid (E330) and candelilla wax (E902).

A face code shows just how user-friendly each additive/ingredient is, from *safe and/or beneficial* through to *hazardous*.

☺☺	**2 happy faces = safe and/or beneficial**
☺	**1 happy face = safe for most people**
☺?	**1 quizzical face = caution advised**
☹	**1 sad face = best avoided**
☹☹	**2 sad faces = hazardous**

A 'traffic lights' system of colour coding: red for 'stop', yellow for 'wait/caution', and green for 'proceed safely' makes reading the tables even simpler.

Note: The evaluation given is the opinion of the author at the time of writing based on available researched information. This information was referenced from several sources, including Material Safety Data Sheets, animal studies, medical and scientific laboratory reports.

The codes shown are only a general guide, as individuals react differently to chemical exposures. The type and severity of reaction will depend on many factors. A few of these are: the health of the person, the amounts to which they are exposed and the period of exposure, the environment in which the person lives/works and the person's age and sex. However, it is recommended that only those products containing additives and ingredients that are *safe and/or beneficial* or *safe for most people* as indicated by the happy faces be chosen.

The tables also show some of the benefits of the additive/ingredient and/or some of the detrimental effects, symptoms and illnesses it has the potential to either cause or exacerbate, and some of the environmental effects that may occur.

In most cases the origin of each additive/ingredient is also given including whether it may be of ANIMAL origin or a product of genetic modification (GM).

Where certain specific medical disorders including cancer, diabetes, tumours and others are mentioned, only limited reference is made as to whether occurrence was in animals or humans. Also, usually no reference is made to the amounts or concentration of chemicals involved, types of exposure or time periods involved. This information is far beyond the scope of this book and the reader is directed to the bibliography if they wish to find out more information.

The tables also list a few relevant common consumer products that may contain that particular additive/ingredient, and some other uses for it.

Beyond the tables, there is also a section on genetic engineering and, for those who want more information than can be included in a book of this size, a list of useful Internet resources.

Disclaimer

Introduction

'Men stumble over the truth from time to time, but most pick themselves up and hurry off as if nothing happened'

Winston Churchill

The idea for this book was born out of a need to understand how some chemicals that are a part of our everyday lives may also play a part in ill health.

Whilst in practice as a homeopathic practitioner I often wondered why some of my patients would regain their health under treatment only to relapse later. It was only after some research that I made the connection between what my patients were eating – not just the types of foods, but also often more significantly the chemical food additives that they contained – and their health problems. I also investigated whether there was a possibility that the products they used on their bodies every day, personal care and cosmetic products, could also have a detrimental effect on their health.

What I discovered during my research amazed and often shocked me. I discovered that a significant number of chemicals added to foods and cosmetics could cause or exacerbate health problems such as asthma, dermatitis, hives, migraines, hay fever, gastric upsets, behavioural problems, hyperactivity, learning difficulties and many others. Some of these chemicals are found to be toxic to body organs and systems like the liver, kidneys, heart, thymus, brain, immune, nervous, hormonal and endocrine systems. Even more disturbing was the fact that some chemicals permitted in foods, personal care products and cosmetics could also cause damage to DNA, birth defects, genetic mutations and cancer.

I began to tell my patients of my discoveries and encouraged them to eliminate, as much as possible, the chemicals that were found to have detrimental effects on health. As I had expected, the health of my patients improved dramatically and often surprised even the patients themselves. An interesting side effect happened as well. Those patients who enrolled their families and friends into this new lifestyle by eliminating harmful chemicals were reporting that the health and well-being of these people was also improving, sometimes dramatically so.

I came to the decision that perhaps I should write a booklet so that many people could benefit from this knowledge. I had envisioned a credit card sized guide that would fit in the wallet. It wasn't long into the research that I realised that the information would overflow such a small format. I ploughed on with my research and after another twelve months felt I had enough information to publish a small shopping guide.

The first edition of this book was self-published as *The Chemical Maze* in Australia in April 2001. Six years later, over 60,000 copies had been sold, mostly in Australia and New Zealand.

Since that book found its way into homes and shopping bags, I have received many letters and emails from people thanking me for producing a user-friendly guide to the vast array of additives and ingredients in our foods, cosmetics and personal care products, and telling me how their health and that of their children has improved after using the book to eliminate harmful chemicals from their lives. I am often touched and inspired by the stories they tell, like the following from Debra Gillis in Sydney:

'I would like to share a miracle with my youngest son Jack. By eliminating harmful chemicals from our house and food, my son's behaviour has become much calmer. The main difference that we welcomed related to Jack's typical ADHD symptoms (not that I ever put him in that box) – this is the first time that anything had actually made a huge difference for him. His memory is getting better each day and he is able to learn more easily than before. His Rudolf Steiner schoolteacher is amazed at the dramatic positive difference since taking Jack off all food colourings, preservatives and away from chemicals.'

This book provides information on over 300 food additives and over 500 of the most common ingredients found in cosmetics and personal care products. Manufacturers can choose from well in excess of 10,000 substances, of which more than 1000 are know to have harmful effects. It's impossible to list all of them in a user-friendly guide, so this book contains those you are most likely to encounter regularly.

A frightening number of the chemicals manufacturers use have never been adequately tested for long-term effects on human health. In addition, there are ingredients in use that some countries have banned.

Looking on the positive side, there are an increasing number of companies producing foods and cosmetic products without harmful E numbers and synthetic chemicals. So we do have choices and with a little bit of guidance and a determination to act we can avoid the nasties and lead healthier lives. The price of ignorance and apathy can be very high indeed. We do not have to pay that price.

Our existence on this planet may well depend on the decisions we make and the actions we take. Now is the time to act!

E Number
Conversion Chart

Manufacturers can identify food additives either by name or by a code known in the United Kingdom and the EU as an E number. The E signifies that the Federation of European Food Additives and Food Enzymes Industries and the European Union have given approval for the use of that additive.

E100	Curcumin
E101	(i) Riboflavin (ii) Riboflavin-5'-phosphate
E102	Tartrazine
E104	Quinoline yellow
E110	Sunset Yellow FCF; Orange Yellow S
E120	Cochineal; Carminic acid; Carmines
E122	Azorubine; Carmoisine
E123	Amaranth
E124	Ponceau 4R; Cochineal Red A
E127	Erythrosine
E128	Red 2G
E129	Allura Red AC
E131	Patent Blue V
E132	Indigotine; Indigo Carmine
E133	Brilliant Blue FCF
E140	Chlorophylls and chlorophyllins
E141	Copper complexes of chlorophyll and chlorophyllins
E142	Green S
E150a	Plain caramel

E150b	Caustic sulphite caramel
E150c	Ammonia caramel
E150d	Sulphite ammonia caramel
E151	Brilliant Black BN; Black PN
E153	Vegetable carbon
E154	Brown FK
E155	Brown HT
E160a	Carotenes
E160b	Annatto; Bixin; Norbixin
E160c	Paprika extract; Capsanthian; Capsorubin
E160d	Lycopene
E160e	Beta-apo-8'-carotenal (C30)
E160f	Ethyl ester of beta-apo-8'-carotenoic acid (C30)
E161b	Lutein
E161g	Canthaxanthin
E162	Beetroot Red; Betanin
E163	Anthocyanins
E170	Calcium carbonates
E171	Titanium dioxide
E172	Iron oxides and hydroxides
E173	Aluminium
E174	Silver
E175	Gold
E180	Litholrubine BK
E200	Sorbic acid
E202	Potassium sorbate
E203	Calcium sorbate
E210	Benzoic acid
E211	Sodium benzoate
E212	Potassium benzoate
E213	Calcium benzoate
E214	Ethyl p-hydroxybenzoate
E215	Sodium ethyl p-hydroxybenzoate

E216	Propyl p-hydroxybenzoate
E217	Sodium propyl p-hydroxybenzoate
E218	Methyl p-hydroxybenzoate
E219	Sodium methyl p-hydroxybenzoate
E220	Sulphur dioxide
E221	Sodium sulphite
E222	Sodium hydrogen sulphite
E223	Sodium metabisulphite
E224	Potassium metabisulphite
E226	Calcium sulphite
E227	Calcium hydrogen sulphite
E228	Potassium hydrogen sulphite
E230	Biphenyl; diphenyl
E231	Orthophenyl phenol
E232	Sodium orthophenyl phenol
E234	Nisin
E235	Natamycin
E239	Hexamethylene tetramine
E242	Dimethyl dicarbonate
E249	Potassium nitrite
E250	Sodium nitrite
E251	Sodium nitrate
E252	Potassium nitrate
E260	Acetic acid
E261	Potassium acetate
E262	Sodium acetate
E263	Calcium acetate
E270	Lactic acid
E280	Propionic acid
E281	Sodium propionate
E282	Calcium propionate
E283	Potassium propionate
E284	Boric acid
E285	Sodium tetraborate; borax
E290	Carbon dioxide

E296	Malic acid
E297	Fumaric acid
E300	Ascorbic acid
E301	Sodium ascorbate
E302	Calcium ascorbate
E304	Fatty acid esters of ascorbic acid
E306	Tocopherols
E307	Alpha-tocopherol
E308	Gamma-tocopherol
E309	Delta-tocopherol
E310	Propyl gallate
E311	Octyl gallate
E312	Dodecyl gallate
E315	Erythorbic acid
E316	Sodium erythorbate
E319	Tertiary butyl hydroquinone
E320	Butylated hydroxyanisole (BHA)
E321	Butylated hydroxytoluene (BHT)
E322	Lecithins
E325	Sodium lactate
E326	Potassium lactate
E327	Calcium lactate
E330	Citric acid
E331	Sodium citrates
E332	Potassium citrates
E333	Calcium citrates
E334	Tartaric acid (L-(+))
E335	Sodium tartrates
E336	Potassium tartrates
E337	Sodium potassium tartrate
E338	Phosphoric acid
E339	Sodium phosphates
E340	Potassium phosphates
E341	Calcium phosphates
E343	Magnesium phosphates

E350	Sodium malates
E351	Potassium malate
E352	Calcium malates
E353	Metatartaric acid
E354	Calcium tartrate
E355	Adipic acid
E356	Sodium adipate
E357	Potassium adipate
E363	Succinic acid
E380	Triammonium citrate
E385	Calcium disodium ethylene diamine tetra-acetate; calcium disodium EDTA
E400	Alginic acid
E401	Sodium alginate
E402	Potassium alginate
E403	Ammonium alginate
E404	Calcium alginate
E405	Propane-1,2-diol alginate
E406	Agar
E407	Carrageenan
E407a	Processed eucheuma seaweed
E410	Locust bean gum; carob gum
E412	Guar gum
E413	Tragacanth
E414	Acacia gum; gum arabic
E415	Xanthan gum
E416	Karaya gum
E417	Tara gum
E418	Gellan gum
E420	(i) Sorbitol
	(ii) Sorbitol syrup
E421	Mannitol
E422	Glycerol
E425	Konjac
E426	Soybean hemicellulose

E431	Polyoxyethylene (40) stearate
E432	Polyoxyethylene sorbitan monolaurate; Polysorbate 20
E433	Polyoxyethylene sorbitan mono-oleate; Polysorbate 80
E434	Polyoxyethylene sorbitan monopalmitate; Polysorbate 40
E435	Polyoxyethylene sorbitan monostearate; Polysorbate 60
E436	Polyoxyethylene sorbitan tristearate; Polysorbate 65
E440	Pectins
E442	Ammonium phosphatides
E444	Sucrose acetate isobutyrate
E445	Glycerol esters of wood rosins
E450	Diphosphates
E451	Triphosphates
E452	Polyphosphates
E459	Beta-cyclodextrin
E460	Cellulose
E461	Methyl cellulose
E462	Ethyl cellulose
E463	Hydroxypropyl cellulose
E464	Hydroxypropyl methyl cellulose
E465	Ethyl methyl cellulose
E466	Carboxy methyl cellulose; Sodium carboxy methyl cellulose
E468	Crosslinked sodium carboxy methyl cellulose
E469	Enzymatically hydrolysed carboxy methyl cellulose
E470a	Sodium, potassium and calcium salts of fatty acids
E470b	Magnesium salts of fatty acids
E471	Mono- and diglycerides of fatty acids

E472a	Acetic acid esters of mono- and diglycerides of fatty acids
E472b	Lactic acid esters of mono- and diglycerides of fatty acids
E472c	Citric acid esters of mono- and diglycerides of fatty acids
E472d	Tartaric acid esters of mono- and diglycerides of fatty acids
E472e	Mono- and diacetyltartaric acid esters of mono- and diglycerides of fatty acids
E472f	Mixed acetic and tartaric acid esters of mono- and diglycerides of fatty acids
E473	Sucrose esters of fatty acids
E474	Sucroglycerides
E475	Polyglycerol esters of fatty acids
E476	Polyglycerol polyricinoleate
E477	Propane-1,2-diol esters of fatty acids
E479b	Thermally oxidised soya bean oil interacted with mono and diglycerides of fatty acids
E481	Sodium stearoyl-2-lactylate
E482	Calcium stearoyl-2-lactylate
E483	Stearyl tartrate
E491	Sorbitan monostearate
E492	Sorbitan tristearate
E493	Sorbitan monolaurate
E494	Sorbitan monooleate
E495	Sorbitan monopalmitate
E500	Sodium carbonates
E501	Potassium carbonates
E503	Ammonium carbonates
E504	Magnesium carbonates
E507	Hydrochloric acid
E508	Potassium chloride
E509	Calcium chloride

E511	Magnesium chloride
E512	Stannous chloride
E513	Sulphuric acid
E514	Sodium sulphates
E515	Potassium sulphates
E516	Calcium sulphate
E517	Ammonium sulphate
E520	Aluminium sulphate
E521	Aluminium sodium sulphate
E522	Aluminium potassium sulphate
E523	Aluminium ammonium sulphate
E524	Sodium hydroxide
E525	Potassium hydroxide
E526	Calcium hydroxide
E527	Ammonium hydroxide
E528	Magnesium hydroxide
E529	Calcium oxide
E530	Magnesium oxide
E535	Sodium ferrocyanide
E536	Potassium ferrocyanide
E538	Calcium ferrocyanide
E541	Sodium aluminium phosphate
E551	Silicon dioxide
E552	Calcium silicate
E553a	(i) Magnesium silicate
	(ii) Magnesium trisilicate
E553b	Talc
E554	Sodium aluminium silicate
E555	Potassium aluminium silicate
E556	Aluminium calcium silicate
E558	Bentonite
E559	Aluminium silicate; Kaolin
E570	Fatty acids
E574	Gluconic acid
E575	Glucono delta-lactone

E576	Sodium gluconate
E577	Potassium gluconate
E578	Calcium gluconate
E579	Ferrous gluconate
E585	Ferrous lactate
E586	4-hexylresorcinol
E620	Glutamic acid
E621	Monosodium glutamate
E622	Monopotassium glutamate
E623	Calcium diglutamate
E624	Monoammonium glutamate
E625	Magnesium diglutamate
E626	Guanylic acid
E627	Disodium guanylate
E628	Dipotassium guanylate
E629	Calcium guanylate
E630	Inosinic acid
E631	Disodium inosinate
E632	Dipotassium inosinate
E633	Calcium inosinate
E634	Calcium 5'-ribonucleotides
E635	Disodium 5'-ribonucleotides
E640	Glycine and its sodium salt
E650	Zinc acetate
E900	Dimethylpolysiloxane
E901	Beeswax, white and yellow
E902	Candelilla wax
E903	Carnauba wax
E904	Shellac
E905	Microcrystalline wax
E912	Montan acid esters
E914	Oxidised Polyethylene wax
E920	L-Cysteine
E927b	Carbamide
E938	Argon

E939	Helium
E941	Nitrogen
E942	Nitrous oxide
E943a	Butane
E943b	Iso-butane
E944	Propane
E948	Oxygen
E949	Hydrogen
E950	Acesulfame K
E951	Aspartame
E952	Cyclamic acid and its Na and Ca salts
E953	Isomalt
E954	Saccharin and its Na, K and Ca salts
E955	Sucralose
E957	Thaumatin
E959	Neohesperidine DC
E962	Salt of aspartame-acesulfame
E965	(i) Maltitol (ii) Maltitol syrup
E966	Lactitol
E967	Xylitol
E968	Erythritol
E999	Quillaia extract
E1103	Invertase
E1105	Lysozyme
E1200	Polydextrose
E1201	Polyvinylpyrrolidone
E1202	Polyvinylpolypyrrolidone
E1204	Pullulan
E1404	Oxidised starch
E1410	Monostarch phosphate
E1412	Distarch phosphate
E1413	Phosphated distarch phosphate
E1414	Acetylated distarch phosphate
E1420	Acetylated starch

E1422	Acetylated distarch adipate
E1440	Hydroxyl propyl starch
E1442	Hydroxy propyl distarch phosphate
E1450	Starch sodium octenyl succinate
E1451	Acetylated oxidised starch
E1452	Starch aluminium octenyl succinate
E1505	Triethyl citrate
E1518	Glyceryl triacetate; triacetin
E1520	Propan-1,2-diol; propylene glycol
E-	Polyethylene glycol 6000

Section 1

Benzyl salicylate (synthetic) — UV ab...

Betaglucans (found in oat fibre and barley) — Thickener

Beta-naphthol (from naphthalene from coal tar) — Solvent

FOOD ADDITIVES

Names	Number	Function	Code	
Acacia gum (gum Arabic; extracted from acacia Senegal)	E414	Thickener Emulsifier	😐?	
Acesulfame K (synthetic chemical)	E950	Artificial sweetener Flavour enhancer	☹	
Acetic acid (occurs naturally in a variety of fruits and plants)	E260	Food acid Buffer	☺	
Acetic acid esters of mono- and diglycerides of fatty acids (may be of ANIMAL origin; may be GM)	E472a	Emulsifier Stabiliser	☺☺	
Acetylated distarch adipate (modified starch; see Starch – Modified in section 2)	E1422	Thickener Stabiliser	😐?	
Acetylated distarch phosphate (synthetic; see Starch – Modified in section 2)	E1414	Thickener Stabiliser	😐?	

	Potential Effects	Possible Food Use	Other Uses
	Low oral toxicity; asthma; skin rash; hives; hay fever	Confectionery, soft drinks, chewing gum, jelly, glazes	Cosmetics, hair products, medicines
	Caused lung tumours, breast tumours, leukaemia, respiratory disease and cancer in animals	Artificial sweetener, low calorie foods, low joule chewing gum	Oral care products
	Regarded as safe in food use; skin irritation; hives; skin rash; caused cancer in rats and mice orally and by injection; harmful to aquatic organisms	Pickles, chutney, cheese, baked goods, sauces	Animal feeds, hair dye, hand lotion, cigarettes
	Regarded as safe in food use	Confectionery, ice cream, bread, dessert toppings, custard mix, cheesecake mix	
	Uncertainties exist about the safety of modified starches especially in infants	Sauces, pickles, yoghurt, dry cake mix, canned fruits, pie fillings	
	Uncertainties exist about the safety of modified starches especially in infants	Sauce, pickles, yoghurt, dry cake mix, canned fruits, pie fillings	

Names	Number	Function	Code
Acetylated oxidised starch (modified starch; see Starch – Modified in section 2)	E1451	Thickener Stabiliser	☺?
Acetylated starch (synthetic; see Starch – Modified in section 2)	E1420	Thickener Vegetable gum	☺?
Adipic acid (prepared from the oxidation of cyclohexanol by nitric acid liberating nitrous oxide, a greenhouse gas)	E355	Raising agent Buffer	☺?
Agar (derived from red algae)	E406	Thickener Emulsifier	☺
Alginic acid (seaweed extract)	E400	Vegetable gum Thickener	☺
Allura Red AC (FDandC Red No40; coal tar dye; banned in some countries)	E129	Colouring (orange/ red)	☹

FOOD ADDITIVES

Potential Effects	Possible Food Use	Other Uses
Uncertainties exist about the safety of modified starches especially in infants	Canned food for infants and young children, confectionery	
Uncertainties exist about the safety of modified starches especially in infants	Sauces, chutney, desserts, baked products, confectionery	
Regarded as safe in food use at low levels; moderately toxic by ingestion; toxic effects in rats including death; teratogenic	Beverages, baked goods, oils, snack foods, processed cheese	Manufacture of plastics and nylons
Regarded as safe in food use at low levels; mildly toxic by ingestion; allergic reactions	Ice cream, baked goods, desserts, manufactured meats, jellies	Cosmetics, bulk laxative
Safe in foods at low levels; alginates inhibited absorption of essential nutrients in some animal tests	Ice cream, dessert mix, custard mix, flavoured milk, cordials, infant formula, yoghurt	Cosmetics and textiles
Asthma; hyperactivity; allergic reactions; hay fever; hives; aspirin sensitive people may wish to avoid; adverse reproductive effects in animals; carcinogenic	Packet cake mix, packet trifle mix, jelly crystals, cereals, chocolate biscuits	Cosmetics, lipstick, medications

Names	Number	Function	Code	
Alpha tocopherol (vitamin E; may be synthetic; may be GM)	E307	Antioxidant Nutrient	☺	
Aluminium (extracted from the mineral ore bauxite)	E173	Colouring (metallic)	☹	
Aluminium ammonium sulphate (made from ammonium sulphate and aluminium sulphate)	E523	Stabiliser Buffer	☺?	
Aluminium calcium silicate	E556	Anticaking agent	☺?	
Aluminium potassium sulphate (made from lime and diatomaceous earth	E552	Buffer Firming agent	☺?	

	Potential Effects	Possible Food Use	Other Uses
	Regarded as safe in food use; may be destroyed by freezing	White flour, white bread, white rice, margarine	
	Ingestion or inhalation can aggravate kidney & lung disorders; cardiovascular, reproductive and neurotoxicity; evidence of a link with Alzheimer's; European Parliament said aluminium additives should be banned	External decoration on cakes etc	Cosmetics, cooking pots and pans, antiperspirant, silver finish to pills and tablets
	Regarded as safe in food use at low levels?; ingestion of large doses can cause burning in mouth and throat, vomiting and diarrhoea; see Aluminium (E173)	Baking powder, milling and cereal industries	Purification of drinking water, fireproofing, vegetable glue
	Regarded as safe in food use at low levels?; see Aluminium (E173)	Garlic salt, table salt, vanilla powder	
	Regarded as safe in foods at low levels?; ingestion of large doses can cause burning in the mouth and throat; see Aluminium (E173)	Cereal, flour, bleached flour, cheese	After-shave lotion, 'size' used for glazing and coating paper

Names	Number	Function	Code	
Aluminium silicate; kaolin	E559	Anticaking agent	☺?	
Aluminium sodium sulphate	E521	Buffer Firming agent	☺?	
Aluminium sulphate	E520	Modifier	☺?	
Amaranth (the synthetic chemical not the grain; FD and C Red No 2; coal tar and azo dye; banned in some countries)	E123	Colouring (bluish red)	☹☹	
Ammonia caramel (may be from sugar beet, sugar cane or maize starch; made using ammonia; may be GM)	E150c	Colouring (dark brown to black) Flavouring	☹	

Potential Effects	Possible Food Use	Other Uses
Regarded as safe in food use?; see Aluminium (E173)	Beer production	Baby powder, bath powder, face powder
Regarded as safe in food use?; skin rash from contact; mild sensitisation; see Aluminium	Cereal, flour, bleached flour, cheese	
Moderately toxic by ingestion; pimples under the arms; allergic reactions; may affect reproduction; see Aluminium (E173)	Sweet and dill pickles, pickle relish, modifier for starch	Antiperspirant, deodorant, skin fresheners, packaging materials
Hyperactivity; hives; asthma; rhinitis; aspirin sensitive people may wish to avoid; may affect reproduction, liver, kidneys, birth defects, carcinogenic; teratogen	Packet cake mix, packet trifle mix, jelly crystals, cereal, soft drinks, blackcurrant products	Lipstick, rouge and other cosmetics
Hyperactivity; may affect liver, stomach, reproduction; caused convulsions in some animal tests; blood toxicity in rats	Soy sauce, oyster sauce, biscuits, jams, dark bread, pickles, chocolate, coatings	

Names	Number	Function	Code	
Ammonium alginate (ammonium salt of alginic acid from seaweed)	E403	Thickener Stabiliser	☺	
Ammonium carbonates	E503	Buffer Neutraliser	☺	
Ammonium hydroxide (prepared from ammonia gas; banned in some countries)	E527	Acidity regulator Neutraliser	☻?	
Ammonium phosphatides (may be synthetic; may be of ANIMAL origin)	E442	Emulsifier Stabiliser	☺☺	
Ammonium sulphate (from ammonia and sulphuric acid)	E517	Flour treatment Stabiliser	☺	
Annatto, bixin, norbixin (annatto is obtained from the annatto tree; bixin and norbixin are extracts)	E160b	Colouring (yellow to pink)	☹	

FOOD ADDITIVES

Potential Effects	Possible Food Use	Other Uses
Alginates may have beneficial effects on health; alginates inhibit absorption of essential nutrients in animal tests	Dessert mix, custard mix, ice cream, yoghurt	Cosmetics; boiler water
Regarded as safe in food use at low levels; contact can cause skin rashes on scalp, forehead and hands	Baking powder, chocolate, cocoa	Permanent wave solution and cream, fire extinguishers
Regarded as safe in food use; irritating to eyes and mucous membranes; hair breakage; toxic by ingestion	Cocoa products, chocolate	Metallic hair dye, barrier cream, stain remover, animal feed, detergent
Regarded as safe in food use	Bread, chocolate, confectionery, frying oils	
Regarded as safe in food use; proved fatal to rats in large doses	Bakery products, dough making	Permanent wave lotion; tanning industry, fertilisers
Annatto may cause irritability and head banging in children; hives; hypotension; pruritis; bixin and norbixin under-going toxicity testing	Margarine, baked goods, reduced fat spreads, dairy products, breakfast cereals	Fabric dye, soap, varnish, body paints

WHAT'S REALLY IN YOUR BASKET?

Names	Number	Function	Code	
Anthocyanins (extracted from grape skins or red cabbage)	E163	Colouring (red/violet)	☺☺	
Argon (an inert gas)	E938	Packaging gas	☺☺	
Ascorbic acid (vitamin C)	E300	Preservative Antioxidant	☺	
Aspartame (prepared from phenyl-alanine and aspartic acid; breaks down to methanol then formalde-hyde in the body; may be GM)	E951	Artificial sweetener Flavour enhancer	☹☹	

FOOD ADDITIVES

	Potential Effects	Possible Food Use	Other Uses
	Regarded as safe in food use; considered to have protective effects in the body	Soft drinks, jams, ice cream, wines, yoghurt, sweets, preserves	Vitamin tablets
	Regarded as safe in food use	Packaging of foods	Welding
	Regarded as safe in foods; vitamin C has many beneficial health effects; excessive consumption can cause skin rashes, painful urination and diarrhoea	Confectionery, breakfast cereals, pressed meats, corned meat	Cosmetic cream, hair dye, hair conditioner
	Headache; depression; anxiety; asthma; fatigue; hyperactivity; MS like symptoms; blindness; aggression; migraine; insomnia; dizziness; irritability; epilepsy; memory loss; seizures - more than 92 in all; NRC; not recommended for women during pregnancy; teratogenic	Artificial sweeteners (Nutrisweet™, Equal™), low calorie foods, diet drinks, chewing gum, soft drinks, instant coffee and may be added to anything which is sugar free or without added sugar	Medications, including those for children

Names	Number	Function	Code	
Azorubine; carmoisine (azo dye; banned in some countries)	E122	Colouring (red)	😦	
Beeswax, white and yellow (obtained from bees; may be synthetic)	E901	Glazing and polishing agent	🙂	
Beetroot red; betanin (extracted from beetroot)	E162	Colouring (deep red/ purple)	🙂	
Bentonite (colloidal clay; aluminium silicate)	E558	Thickener Anticaking agent	😐?	
Benzoic acid (occurs in nature in cherry bark, raspberries, anise and cassia bark; may be made commercially from benzene)	E210	Preservative Flavouring	😦	
Beta-apo-8'-carotenal (C30) (synthetic)	E160e	Colouring (orange to yellow/red)	🙂🙂	

	Potential Effects	Possible Food Use	Other Uses
	Asthma; hyperactivity; aspirin sensitive people may wish to avoid; animal carcinogen	Confectionery, sweets, Marzipan, brown sauce, jelly crystals	
	Regarded as safe in food use; can cause mild allergic reactions and contact dermatitis	Confectionery, soft drinks, chewing gum	Mascara, eye make-up, baby cream, lipstick, cosmetics
	Regarded as safe in food use; contains nitrates so NRC	Desserts, jellies, jams, liquorice, sweets	Cosmetics
	Regarded as safe in food use; may clog skin pores inhibiting proper skin function; see Aluminium (E173)	Colouring in wine, sugar brewing and purification, settling wine sediments	Cosmetics, facial masks, animal and poultry feeds, detergents
	Asthma; hives; behavioural problems; hyperactivity; may affect lungs; eye and skin irritation; aspirin sensitive people may wish to avoid; neurotoxicity	Brewed soft drinks, cider, non-dairy dip, chewing gum, fruit juice, margarine, ice cream	Cosmetics, hair rinse, skin cleanser, perfume, pharmaceuticals
	Regarded as safe in food use.	Cream cheese spread, cheese slices, processed cheese	

Names	Number	Function	Code	
Beta-cyclodextrin (BCD; naturally occurring from the action of enzymes on starch)	E459	Processing aid Stabiliser	☺☺	
BHA	E320	Antioxidant Preservative	☹☹	
BHT	E321	Antioxidant Preservative	☹☹	
Biphenyl; diphenyl (made from benzene; banned in some countries)	E230	Antifungal Preservative	☹☹	
Boric acid (made by the action of sulphuric or hydrochloric acid on borax; on Canadian Hotlist)	E284	Preservative Oral care agent	☹☹	
Brilliant black BN; black PN (azo dye; banned in many countries)	E151	Colouring (black)	☹☹	

	Potential Effects	Possible Food Use	Other Uses
	Regarded as safe in food use	Processed foods	Toothpaste, skin cream
	See butylated hydroxyanisole		
	See butylated hydroxyanisole		
	Exposure can cause eye and nasal irritation; vomiting; nausea; CNS depression; liver, kidney, respiratory, cardiovascular and neurotoxicity	Used to fumigate some fruits (residue on fruit skins), food wrapped in paper impregnated with diphenyl, citrus peel, marmalade	
	Severe poisonings have occurred after ingestion or application to abraded skin; kidney, cardiovascular, reproductive, liver and neurotoxicity	Caviar, fungus control on citrus fruit	Baby powder, bath powder, soap, eye cream, mouthwash, cosmetics
	Asthma; hyperactivity; may affect kidneys, stomach; NRC; carcinogenic	Blackcurrant cake mix, brown sauce, eggnog, drinking yoghurt	

Names	Number	Function	Code	
Brilliant blue FCF (FD and C Blue No1; banned in some countries)	E133	Colouring (bright blue)	☹☹	
Brown FK (mixture of six azo dye, other colours plus sodium chloride and/ or sodium sulphate; banned in many countries)	E154	Colouring (brown)	☹☹	
Brown HT (coal tar and azo dye; banned in some countries)	E155	Colouring (brown)	☹	
Butane (petroleum derivative)	E943a	Solvent Propellant	☺?	
Butylated hydroxyanisole (BHA; petroleum derivative; banned in some countries)	E320	Antioxidant Preservative	☹☹	

Potential Effects	Possible Food Use	Other Uses
Asthma; hives; hay fever; allergic reactions; NRC; aspirin sensitive people may wish to avoid; carcinogenic	Gelatine, canned processed peas, soft drinks, dairy products, cereals, desserts	Toothpaste, cosmetics, hair dye, deodorant
Asthma; allergic reactions; hives; may affect heart, kidneys, liver; thyroid; aspirin sensitive people may wish to avoid; NRC	Kippers, smoked and cured fish, cooked ham, potato chips	
Asthma; allergic reactions; hives; may affect kidneys, NRC; aspirin sensitive people may wish to avoid	Chocolate cake mix, chocolate biscuits	
Animal carcinogen; CNS depression; on NIH hazards list; neurotoxicity	Refrigerant	Aerosol cosmetics
Hives; somnolence; hay fever; headache; wheezing; fatigue; asthma; NRC; may affect kidneys, liver, thyroid, stomach, re-production; hormone disruption; carcinogenic; teratogen	Instant mashed potato, edible fats and oils, chewing gum, reduced fat spread, margarine, processed meats, ice cream	Cosmetics

Names	Number	Function	Code	
Butylated hydroxytoluene (BHT; petroleum derivative: banned in some countries)	E321	Antioxidant Preservative	☹☹	
Calcium acetate	E263	Food acid Firming agent	☺	
Calcium alginate	E404	Thickener Stabiliser	☺	
Calcium ascorbate (prepared from ascorbic acid and calcium carbonate)	E302	Preservative Antioxidant	☺☺	
Calcium benzoate (calcium salt of benzoic acid)	E213	Preservative	☹	

FOOD ADDITIVES

	Potential Effects	Possible Food Use	Other Uses
	Chronic hives; dermatitis; fatigue; asthma; aggressive behaviour; bronchospasm; NRC; may affect reproduction, kidneys, stomach, liver; harmful to aquatic organisms	Edible fats and oils, chewing gum, fish products, dry breakfast cereals, beer & malt drinks, polyethylene film for wrapping food	Shaving cream, baby oil, baby lotion, lipstick, eyeliner, packaging materials, rubber, jet fuel
	Regarded as safe in food use; low oral toxicity	Bread, pickles, beer, ale, cheese, salad cream	Dyeing and curing skins, cosmetic fragrance
	Alginates may have beneficial effects on health; they inhibited absorption of essential nutrients in animal tests	Ice cream, soft and cottage cheeses, cheese snacks, instant desserts	Hand lotion and cream, shampoo, wave sets
	Regarded as safe in food use at low levels	Concentrated milk products, cooked and cured meat products	
	Asthma; hives; anaphylaxis; hyperactivity; behavioural problems; eczema; aspirin sensitive people may wish to avoid; NRC	Brewed soft drinks, non-dairy dip, chewing gum, fruit juice, margarine, ice cream	

Names	Number	Function	Code	
Calcium carbonate (chalk, limestone, marble, dolomite, coral)	E170	Colouring (white) Firming agent	☺	
Calcium chloride (chloride salt of calcium)	E509	Firming agent Sequestrant	☺	
Calcium citrates (prepared from citrus fruits)	E333	Food acid Buffer	☺	
Calcium diglutamate (calcium salt of glutamic acid; contains MSG; may be GM)	E623	Flavour enhancer	☺?	
Calcium disodium ethylene diamine tetra-acetate; calcium disodium EDTA (banned in some countries)	E385	Preservative Sequestrant	☹	

Potential Effects	Possible Food Use	Other Uses
Regarded as safe in food use at low levels; excess can cause abdominal pain; constipation	Bread, biscuits, confectionery, ice cream, cakes, sweets, canned fruit and vegetables	Cosmetics, face powder, bleaches, vitamin tablets, cigarettes
Regarded as safe in food use at low levels; irritation of skin and mucous membranes; stomach upsets; irregular heartbeat	Cottage cheese, jellies, canned tomatoes, low sodium salt substitute	Cosmetics, eye lotion, fire extinguishers
May provoke symptoms in those who react to MSG; citrates may interfere with the results of laboratory tests for blood, liver and pancreatic function	Confectionery, jellies, jams, to improve baking properties in flour	
Asthma; aspirin sensitive people may wish to avoid; probably similar to MSG	Low sodium salt substitute	
Muscle cramps; blood in the urine; intestinal upset; kidney damage; mineral imbalance; chromosome damage; can increase uptake of heavy metals; may affect liver and reproduction	Dressings, soft drinks, sandwich spreads, beer, ale, margarine, instant teas	Used medically to detoxify heavy metal poisoning

Names	Number	Function	Code	
Calcium ferrocyanide (synthetic)	E538	Anticaking agent	☺☺	
Calcium gluconate (made from calcium carbonate and gluconic acid)	E578	Buffer Sequestrant	☹?	
Calcium guanylate	E629	Flavour enhancer	☺	
Calcium hydrogen sulphite (see Sulphites section 2)	E227	Preservative Firming agent	☹	
Calcium hydroxide (slaked lime)	E526	Acidity regulator Firming agent	☺	
Calcium inosinate (calcium salt of inosinic acid; of ANIMAL origin)	E633	Flavour enhancer	☺	
Calcium lactate (may be of ANIMAL origin)	E327	Food acid Buffer	☹?	
Calcium malates (calcium salts of malic acid)	E352	Buffer Firming agent	☺☺	

	Potential Effects	Possible Food Use	Other Uses
	Regarded as safe in food use	Table salt	
	Gastric irritation; stomach problems, heart problems	Preserves, infant formula, anticaking of coffee powder	Cosmetics, animal feed
	Regarded as safe in food use; may trigger gout	Processed foods, condiments and seasonings	
	Asthma; gastric irritation; may affect liver, kidneys, lungs, stomach	Canned fruits and vegetables	
	Regarded as safe in food use; toxic and hazardous in concentrated form	Canned peas, fruit products, infant formula, beer, ale	Depilatories, animal feeds, plaster, pesticides
	Regarded as safe in food use; may trigger gout	Processed foods, condiments and seasonings	
	Regarded as safe in food use; may cause cardiac and gastro-intestinal disturbance; people with lactose intolerance may wish to avoid	Confectionery, baking powder, canned bean sprouts, condensed milk	Oral menstrual drug products? (unsafe), dentifrices, animal feeds
	Regarded as safe in food use	Fruit drinks, soft drinks, sweetened coconut	

Names	Number	Function	Code	
Calcium oxide (quick lime; strongly caustic)	E529	Emulsifier Texturiser	😐?	
Calcium phosphates (from phosphoric acid)	E341	Buffer Sequestrant	☺	
Calcium propionate (calcium salt of propionic acid)	E282	Mould inhibitor Preservative	😐?	
Calcium 5'-ribonucleotides (may be of ANIMAL origin)	E634	Flavour enhancer	☹	
Calcium silicate (made from lime and diatomaceous earth)	E552	Anticaking agent Glazing agent	☺	

Potential Effects	Possible Food Use	Other Uses
Regarded as safe in food use; can cause severe damage to skin and mucous membranes on contact; thermal and chemical burns	Confectionery, custard mix, flour products, soup, malted milk powder, canned peas	Cosmetics, home and garden pesticides, insecticides, plaster
Regarded as safe in food use; can cause skin and eye irritation on contact	Flour products, malted milk powder, cereal flours, condiments	Toothpaste, tooth powder, cosmetics
Irritability; asthma; migraine; fatigue; learning difficulties; aggression; gastric irritation; headaches; sensitivity to propionates occurs in conjunction with sensitivity to other chemicals	Bread, processed cheese, poultry stuffing, chocolate products	Cosmetics, antifungal medication
Asthma; hyperactivity; itchy skin rashes up to 30 hrs after eating, swelling of lips, throat and tongue; anaphylaxis	Flavoured crisps, instant noodles, manufactured pies	
Regarded as safe in food use; inhalation may cause respiratory tract irritation; asthma	Baking powder, rice, chewing gum, table salt, vanilla powder	Face powder, lime glass, cement

Names	Number	Function	Code	
Calcium sorbate (synthetic; derived from sorbic acid)	E203	Preservative	☹?	
Calcium stearoyl-2-lactylate (calcium salt of lactyl lactate; may be of ANIMAL origin)	E482	Emulsifier Stabiliser	☺	
Calcium sulphate (becomes Plaster of Paris when heated)	E516	Firming agent	☺	
Calcium sulphite (a salt of sulphurous acid; see Sulphites in section 2)	E226	Preservative Firming agent	☹	
Calcium tartrate (derived from cream of tartar)	E354	Acidity regulator Sequestrant	☺☺	
Candelilla wax (from the candelilla plant)	E902	Glazing agent Emollient	☺	

Potential Effects	Possible Food Use	Other Uses
Contact hives; skin irritation; asthma; allergic reactions; behavioural problems	Bread, cheese spread, cottage cheese, soft drinks, chocolate syrup, cheese-cake	Ointments, cosmetics
Regarded as safe in food use; adverse reactions have occurred in animals during testing	Flour for making bread, biscuits, instant mashed potatoes, processed egg whites	Powdered cosmetics
Regarded as safe in food use; large amounts may cause intestinal obstruction and constipation	Flour products, baking powder, cereal flours, canned tomatoes, Gorgonzola cheese	Toothpastes and powder, brewing industry, insecticides
Asthma; skin irritation; gastric irritation; may affect kidneys, gastro-intestinal tract and liver	Canned fruits and vegetables	Disinfectant, home wine brewing
Regarded as safe in food use	Biscuits, rusks	Tobacco
Regarded as safe in food use but needs further testing	Coating for foods, chewing gum	Lipstick, writing inks, cosmetics

Names	Number	Function	Code	
Canthaxanthin (usually from beta carotene but may be of ANIMAL origin; banned in some countries)	E161g	Colouring (pink)	☺?	
Carbamide (urea; may be of ANIMAL origin)	E927b	Browning agent Deodoriser	☺	
Carbon dioxide (commercially produced by fermentation)	E290	Packaging gas Preservative	☺	
Carboxy methyl cellulose; sodium carboxy methyl cellulose (made from cotton by-products; may be GM)	E466	Thickener Stabiliser	☺?	
Carnauba wax (from a Brazilian wax palm tree)	E903	Glazing agent Texturiser	☺	
Carotenes (mostly natural and of plant origin; may be of ANIMAL origin)	E160a	Colouring (orange to red)	☺☺	

Potential Effects	Possible Food Use	Other Uses
Loss of night vision; skin discolouration; sensitivity to glare; recurrent hives; 'gold dust' retinopathy	Fish fingers, ice cream, mallow biscuits, pickles, sauces, preserves	
Regarded as safe in food use; skin irritation; allergic reactions; headache	Baked goods, chewing gum, pretzels	Roll on deodorant, shampoo, mouthwash, hair colouring
Probably safe with food use; may reduce fertility; teratogenic; neurotoxicity	Confectionery, carbonated beverage, gassed cream	Dry ice, stage fog or smoke effects
Poorly absorbed; flatulence; large amounts can cause diarrhoea and abdominal cramps; caused cancer and tumours in animal studies	Infant formula, ice cream, icings, confectionery, cottage cheese, cream cheese spread	Hair setting lotion, hand cream, medication, laxatives, antacids, tobacco
Rarely causes allergic reactions; contact dermatitis; gastric irritation	Confectionery, waxed fruit, fruit juice, sauces	Cosmetics, lipstick, mascara, varnishes
Regarded as safe in food use at low levels	Margarine, dairy blend, reduced fat spread, cakes, jams, cheese	Cosmetics, animal feed, cigarettes

Names	Number	Function	Code	
Carrageenan (may be degraded, un-degraded or native; may contain or create MSG)	E407	Thickener Vegetable gum	☹	
Caustic sulphite caramel (may be from sugar beet, sugar cane or maize starch; may be GM)	E150b	Colouring (dark brown to black) Flavouring	☺?	
Cellulose (prepared from wood pulp; may be GM)	E460	Anticaking agent Binder	☺☺	
Chlorophylls and chlorophyllins (green colouring in plants)	E140	Colouring (olive to dark green)	☺	
Citric acid (derived from citrus fruit or corn; often contains MSG; may be GM, see also Alpha hydroxy acids in section2)	E330	Acidity regulator Flavouring	☺?	

Potential Effects	Possible Food Use	Other Uses
May affect gastrointestinal tract; stomach; NRC; ulcerative colitis; animal carcinogen	Ice cream, dessert mix, confectionery, pastries, biscuits, chocolate products	Cosmetics, cough medicines, toothpaste
Hyperactivity; may affect liver; stomach problems	Whisky, brandy, ice cream	
Regarded as safe in food use	Cakes, sauces, soups, biscuits, bread, spreads, jams, ice cream	Cosmetic cream, tablets
Regarded as safe in food use; can cause a sensitivity to light	Soups, sauces, olive oil, soybean oil	Antiperspirant, deodorant, mouthwash
Regarded as safe in foods; has a number of health benefits; may provoke symptoms in those who react to MSG; may aggravate the herpes simplex virus; in cosmetic use it may cause exfoliative dermatitis; eye and skin irritation	Biscuits, cheese, ice cream, jams, jellies, processed cheese, soft drinks, fruit drinks, infant formula	Freckle cream, eye lotion, nail bleaches, skin fresheners, hair rinses

Names	Number	Function	Code	
Citric acid esters of mono- and diglycerides of fatty acids (may be of ANIMAL origin)	E472c	Emulsifier	☺	
Cochineal; carminic acid; carmines (of ANIMAL origin; cochineal and carminic acid restricted in some countries)	E120	Colouring (red)	☹	
Copper complexes of chlorophyll and chlorophyllins (may be synthetic)	E141	Colouring (bright green)	☺☺	
Crosslinked sodium carboxymethyl-cellulose	E468	Carrier	☻?	
Curcumin (derived from turmeric)	E100	Colouring (orange/yellow) Antioxidant	☺	
Cyclamic acid and its Na and Ca salts (banned in some countries)	E952	Artificial Sweetener Flavour enhancer	☹	

Potential Effects	Possible Food Use	Other Uses
May provoke symptoms in those who react to MSG	Infant formula, foods for infants and young children	
Asthma; anaphylaxis (possibly life-threatening); hay fever; hives; aspirin sensitive people may wish to avoid	Some alcoholic drinks, red applesauce, pie fillings, meats, baked goods, yoghurt	Cosmetics, red eye make-up, shampoo, mascara
Regarded as safe in food use	Soup, sauce, natural fruits in liquid	
See Carboxy Methyl -cellulose (E466)	Sweeteners	
Has many beneficial health effects; may cause skin irritation; moderately toxic by injection	Curry powder, fish fingers, margarine, confectionery, processed cheese, savoury rice	
Migraines; various skin conditions; hives; pruritis; may affect kidneys, heart, circulation, blood, reproduction, liver, thyroid; carcinogen	Artificially sweetened canned fruit, brewed soft drinks, low calorie foods	

Names	Number	Function	Code	
Delta tocopherol (vitamin E; may be GM)	E309	Antioxidant	☺☺	
Dimethyl dicarbonate	E242	Preservative Fungicide	☺☺	
Dimethylpolysiloxane	E900	Antifoaming agent	☺?	
Diphosphates	E450	Buffer Sequestrant	☺	
Dipotassium guanylate (from guanylic acid)	E628	Flavour enhancer	☺	
Dipotassium inosinate (of ANIMAL origin)	E632	Flavour enhancer	☺	
Disodium guanylate (often used in combination with MSG)	E627	Flavour enhancer	☺?	

	Potential Effects	Possible Food Use	Other Uses
	Vitamin E has been shown to play protective roles in the body	Salad oil, reduced fat spread, dairy blend, margarine	
	Regarded as safe in food use	Sport drinks, fruit drinks, instant teas	Yeast inhibitor in wine
	Acute or delayed hypersensitivity reactions; nausea; diarrhoea	Chewing gum, soft drinks, jam, instant coffee, cordials	Ointment base, topical drugs, skin protectant
	Regarded as safe in food use at low levels; excess may cause kidney damage, decrease in bone density; osteoporosis	Canned meat and fish, processed cheese, baking powder	
	Regarded as safe in food use; people with gout or uric acid kidney stones may wish to avoid	Canned vegetables	
	Regarded as safe in food use; may trigger gout	Processed foods, condiments and seasonings	
	Aspirin sensitive people may wish to avoid; people with gout or uric acid kidney stones may wish to avoid; NRC	Canned foods, sauces, snack foods, soups	

Names	Number	Function	Code	
Disodium inosinate (of ANIMAL origin; often contains MSG)	E631	Flavour enhancer	☺?	
Disodium 5'-ribonucleotides (may be of ANIMAL origin; banned in some countries)	E635	Flavour enhancer	☹	
Distarch phosphate (see Starch – Modified in section 2)	E1412	Thickener Stabiliser	☺?	
Dodecyl gallate (ester of gallic acid derived from tannin)	E312	Antioxidant	☺?	
Enzymatically hydrolysed carboxy methylcellulose	E469	Thickener Stabiliser	☹	

Potential Effects	Possible Food Use	Other Uses
People with gout or uric acid kidney stones may wish to avoid; kidney problems; NRC	Canned vegetables	
Asthma; hyperactivity; mood changes; itchy skin rashes up to 30 hrs after consuming; aspirin sensitive people may wish to avoid; not permitted in foods for babies; NRC; gout; kidney problems	Flavoured crisps, instant noodles, manufactured pies	
Uncertainties exist about the safety of modified starches especially in infants	Fruit pie fillings, instant puddings, instant desserts, sauces, soup mix	Cosmetics, water softener
Allergic reactions; NRC; contact dermatitis; gastric irritation; aspirin sensitive people may wish to avoid; caused pathological changes in the spleen, kidneys and liver in test rats	Dairy blend, edible fats and oils, reduced fat spread, margarine	Cosmetic cream, ink
See Carboxy Methyl cellulose (E466)	Ice cream, soup, fatty meat products	

Names	Number	Function	Code	
Erythorbic acid (produced from sugar)	E315	Antioxidant Preservative	☺	
Erythritol (produced by fermentation from glucose)	E968	Sweetener Humectant	☺	
Erythrosine (FD&C Red No3; coal tar dye; banned in some countries)	E127	Colouring (bluish pink)	☹☹	
Ethyl cellulose (chemically prepared from wood pulp or chemical cotton)	E462	Emulsifier Binder	☺	
Ethyl ester of beta-apo-8'-carotenoic acid(C30) (natural substance from plants)	E160f	Colouring (yellow to orange)	☺☺	

	Potential Effects	Possible Food Use	Other Uses
	May cause allergic reactions in some people; has only 5% of the vitamin capacity of ascorbic acid	Breakfast cereal, beverages, flour products, confectionery, pressed meat products	Cosmetics
	Regarded as safe in food use; excess may have a laxative effect	Beverages, baked goods, chocolate, processed foods, liqueurs	Moisturising creams and lotions
	Asthma; hyperactivity; hives; learning difficulties; light sensitivity; may affect liver, heart, thyroid, reproduction, stomach; carcinogenic	Canned fruit cocktail, biscuits, glace cherries, packet trifle mix, canned red cherries, maraschino cherries, sausage casings	Toothpaste, dental disclosing tablets, rouge, medications
	Regarded as safe in food use; excess may cause gastrointestinal upset	Confectionery, chewing gum	Nail polish, lip rouge, cigarettes
	Regarded as safe in food use	Processed foods	

Names	Number	Function	Code	
Ethyl methyl cellulose (from wood pulp or chemical cotton)	E465	Thickener Emulsifier	😐?	
Ethyl p-hydroxybenzoate (ethylparaben; see Parabens in section 2)	E214	Preservative	☹	
Fatty acid esters of ascorbic acid (may be of ANIMAL origin)	E304	Antioxidant Preservative	☺☺	
Fatty acids (may be of ANIMAL origin; may be GM)	E570	Emulsifier Binder	☺	
Ferrous gluconate (made from barium gluconate and ferrous sulphate; may be GM)	E579	Colour retention agent Flavouring	☹	

	Potential Effects	Possible Food Use	Other Uses
	Can cause digestive problems; diarrhoea; flatulence; gastrointestinal disturbances	Vegetable fat, low fat cream, imitation ice cream, whipped toppings	Bulk laxative, tobacco
	Asthma; hives; allergic reactions; skin redness, itching and swelling; anaphylaxis	Brewed soft drinks, ice cream, cider, non-dairy dip, chewing gum, fruit juice, margarine	Cosmetic products
	Regarded as safe in food use	Dairy blend, salad oil, instant mashed potato, reduced fat spread	
	Regarded as safe in food use; contact may cause skin irritation; allergic reactions	Essences, soft drinks, artificial sweeteners, fruit flavoured drinks	Bar soap, lipstick, bubble bath, lubricants, detergent
	Toxic in large amounts; diarrhoea; vomiting; may affect gastrointestinal tract, liver, stomach; caused tumours in mice	Infant formula, formula dietary food, preserved ripe olives	Iron supplements

Names	Number	Function	Code	
Ferrous lactate (may be of ANIMAL origin)	E585	Colour fixative	☺	
Fumaric acid (made by the fermentation of glucose or molasses by fungi)	E297	Food acid Antioxidant	☺☺	
Gamma-tocopherol (synthetic; may be GM)	E308	Antioxidant	☺☺	
Gellan gum (gum made by the fermentation of a carbohydrate with *pseudomonas elodea*)	E418	Thickener Stabiliser	☺	
Gluconic acid (made synthetically from corn; may be GM)	E574	Anticaking agent Sequestrant	☺☺	
Glucono delta-lactone (made from the oxidation of glucose; may be GM)	E575	Acidity regulator	☺	

	Potential Effects	Possible Food Use	Other Uses
	Regarded as safe in food use; caused tumours in mice when injected under the skin	Dietary supplement	
	Regarded as safe in food use	Brewed soft drinks, confections, packet cheesecake mix	Cosmetics
	This type of Vitamin E has been shown to play several protective roles in the body	Dairy blend, salad oil, margarine, reduced fat spread	
	Regarded as safe in food use at low levels; excess can cause diarrhoea	Various foods	
	Regarded as safe in food use	Unstandardised foods	Cosmetics, metal cleaners and degreasers
	Regarded as safe in food use at low levels; excess can cause diarrhoea	Cottage cheese, meat processing, jelly powder, canned vegetables	Cleaning agents, brewing beer

Names	Number	Function	Code	
Glutamic acid (synthetically made from vegetable protein; contains MSG; may be GM)	E620	Flavour enhancer Antioxidant	☺?	
Glycerol (glycerin; synthetic; by-product of soap manufacture; may be of ANIMAL origin)	E422	Humectant Solvent	☺	
Glycerol esters of wood rosins (made from wood rosin and food grade glycerin; may be of ANIMAL origin)	E445	Emulsifier Stabiliser	☺?	
Glyceryl triacetate; triacetin	E1518	Solvent	☺	
Glycine and its sodium salt (may be synthetic; may be of ANIMAL origin)	E640	Flavour enhancer	☺	

Potential Effects	Possible Food Use	Other Uses
Asthma; headache; nausea; sleep disturbances	Adding meat flavour to foods, improving the taste of beer (together with hydrochloric acid	Cosmetics, permanent wave lotion, treatment of epilepsy
Regarded as safe in food use at low levels; mental confusion; headache; may affect stomach, heart, reproduction, blood sugar levels	Confectionery, dried fruit, low calorie foods, marshmallows, baked goods, chewing gum	Tobacco, soap, toothpaste, hand cream, mouthwash, barrier cream, perfumery
Not granted GRAS status in the USA due to insufficient safety data	Chewing gum base, flavouring oils in beverages	
Regarded as safe in food use; high dose injections fatal to rats	Coating for vegetables and fruits	Hair dye, toothpaste, cigarette filters, perfumery
Mildly toxic by ingestion	Used to mask the aftertaste of saccharin	Dietary supplement, cosmetics, antacid

Names	Number	Function	Code	
Gold (naturally occurring metal)	E175	Colouring (metallic)	☺	
Green S (coal tar dye; banned in some countries)	E142	Colouring (green)	☹	
Guanylic acid (made mainly from yeast; often combined with MSG)	E626	Flavour enhancer	☺	
Guar gum (obtained from the seeds of a tree in India)	E412	Thickener Stabiliser	☺	
Helium (from natural gas)	E939	Packaging gas Propellant	☺☺	
Hexamethylene tetramine (derivative of benzene; breaks down to formaldehyde and ammonia; banned in many countries)	E239	Preservative	☹	

	Potential Effects	Possible Food Use	Other Uses
	Regarded as safe in food use; rare allergic reactions; neurotoxicity	External decoration on chocolate confectionery	Cosmetics
	Hyperactivity; asthma; skin rashes; insomnia; see Coal Tar in section 2	Canned peas, mint sauce, packet cheesecake mix	Textile industry
	Mildly toxic by ingestion	Canned foods, sauces, soup, snacks	
	Regarded as safe in food use at low levels; excess can cause abdominal cramps; nausea; flatulence; diarrhoea	Baked goods, jam, cereals, cheese spreads, jellies, beverages, infant foods, toppings	Binding tablets, cosmetics, slimming aids (caution recommended)
	Regarded as safe in food use		Arc welding, inflating balloons
	Contact dermatitis; gastrointestinal upset; kidney damage; genetic mutation in animals; suspected carcinogen	Provolone cheese	Cosmetics, adhesives, coatings lubricating oils

Names	Number	Function	Code	
4-Hexylresorcinol (resorcinol is obtained from various resins)	E586	Antioxidant Anti-browning agent	☹	
Hydrochloric acid	E507	Acidity regulator Modifier	☺☺	
Hydrogen	E949	Packaging gas	☺☺	
Hydroxy propyl distarch phosphate (see Starch – Modified in section 2; may be GM)	E1442	Thickener Emulsifier	☺?	
Hydroxypropyl starch (see Starch – Modified in section 2)	E1440	Thickener Vegetable gum	☺?	
Hydroxypropyl cellulose (synthetic ether of cellulose; may be GM)	E463	Thickener Emulsifier	☺	
Hydroxypropyl methylcellulose (synthetic; from cellulose; may be GM)	E464	Emulsifier Thickener	☺	

	Potential Effects	Possible Food Use	Other Uses
	Large amounts can cause severe gastrointestinal irritation; bowel, liver and heart damage	Uncooked crustaceans such as shrimps, prawns and lobsters	Mouthwash, soap, throat lozenges, sunburn cream
	Regarded as safe in food use	Cottage cheese, cream cheese	Hair bleach, solvent
	Regarded as safe in food use	Hydrogenation of oils like soybean, corn and cottonseed	Production of sorbitol used in cosmetics etc
	Uncertainties exist about the safety of modified starches especially in infants	Canned soups, frozen desserts, sauces, cake mix	
	Uncertainties exist about the safety of modified starches especially in infants	Sauces, pickles, yoghurt, dry cake mix, canned fruit pie fillings	
	Regarded as safe in food use; may cause allergic reactions	Low fat cream, UHT cream	Cosmetics, tobacco
	Regarded as safe in food use; mild eye and skin irritation; allergic reactions	Confectionery, infant formula, icing, topping, ice cream, pickles, soup, dried mixed 'fruit'	Cosmetics, hair and skin preparations, bubble bath, tanning lotion

Names	Number	Function	Code	
Indigotine; indigo carmine (FDandC Blue No2; coal tar dye; banned in some countries)	E132	Colouring (moderate bright green)	☹	
Inosinic acid (from meat extract or dried sardines; of ANIMAL origin)	E630	Flavour enhancer	☺	
Invertase (from the fermentation of yeast; probably GM)	E1103	Processing aid	☺?	
Iron oxides and hydroxides (rust; synthetic oxides and hydroxides of iron; banned in some countries)	E172	Colouring (red/brown/black/orange/yellow)	☺?	
Iso-butane (petroleum derivative)	E943b	Propellant	☺?	
Isomalt (produced from sugar; may be enhanced with Acesulphame K (E950)	E953	Artificial sweetener Thickener	☺	

Potential Effects	Possible Food Use	Other Uses
Asthma; allergic reactions; hyperactivity; heart problems; NRC; carcinogenic; see Coal Tar in section 2	Bottled soft drinks, sweets, biscuits, confectionery, ice cream, bakery products	Hair rinses, dye in kidney function tests, tablets and capsules
Regarded as safe in food use; may trigger gout	Processed foods, condiments and seasonings	
GM enzymes have been linked to serious health concerns and even death	Confectionery with soft centres	
Iron is potentially toxic in all forms; excess can lead to increased risk of numerous health conditions	Salmon and shrimp paste or spread, cake and dessert mixes, meat paste	Pet foods, dying egg shells, face powder, eye make-up
See Butane	Spray-on pan coatings	Cosmetic spray, refrigeration
Regarded as safe in food use; gastric irritation	Ice cream, jams, baked goods	Cosmetics

Names	Number	Function	Code	
Karaya gum (exudate of a tree found in India)	E416	Thickener Stabiliser	☺?	
Konjac (derived from the tubers of a plant grown in Japan)	E425	Emulsifier Thickener	☺?	
Lactic acid (produced commercially from whey, cornstarch, potatoes and molasses; may be of ANIMAL origin; may be GM)	E270	Food acid Preservative	☺	
Lactic acid esters of mono-and diglyc-erides of fatty acids (may be of ANIMAL origin)	E472b	Emulsifier Stabiliser	☺☺	
Lactitol (derived from milk sugar (lactose); of ANIMAL origin)	E966	Artificial sweetener Texturiser	☺☺	
L-cysteine (manufactured from animal hair and chicken feathers; of ANIMAL origin)	E920	Flavouring Improving agent	☺	

	Potential Effects	Possible Food Use	Other Uses
	Asthma; hives; hay fever; dermatitis; reduces nutrient intake; gastric irritation	Ice cream, baked goods, sweets, gumdrops, frozen dairy desserts	Hair spray, setting lotion, hand lotion, toothpaste, shaving cream
	Choking hazard in dry form; diarrhoea; abdominal pain; stomach problems; nutrient disruption	Soups, gravy, jam, mayonnaise, lollies	
	Regarded as safe in food use. In cosmetics stinging of the skin in sensitive people; not recommended for babies under 3 months	Infant formula, salad dressings, confectionery, soft drinks, tartare sauce	Cosmetics, skin fresheners, cigarettes
	Regarded as safe in food use	Bakery products, ice cream, peanut butter, whipped toppings	
	Regarded as safe in food use	Baked goods, chewing gum	
	Has beneficial health effects; may provoke symptoms in those who react to MSG	Flour for bread making, chicken flavouring	Shampoo

Names	Number	Function	Code	
Lecithins (may be from soybean; may be GM)	E322	Stabiliser Emulsifier	☺	
Litholrubine BK (synthetic; azo dye; banned in some countries)	E180	Colouring (reddish)	☹	
Locust bean gum; carob gum	E410	Thickener Emulsifier	☺	
Lutein (found in egg yoke, fat cells and green leaves; may be of ANIMAL origin)	E161b	Colouring (yellow to red)	☺☺	
Lycopene (extracted from tomatoes and pink grapefruit; may be synthetic; may be GM)	E160d	Colouring (red)	☺☺	
Lysozyme (egg protein; of ANIMAL origin; may be GM)	E1105	Preservative	☺?	

Potential Effects	Possible Food Use	Other Uses
Regarded as safe in food use; people with allergy to soy may wish to avoid	Chocolate, dried milk, margarine, dessert mix, confectionery	Cosmetics, lipstick, hand cream, pharmaceuticals
Asthma; hives; hay fever; gastrointestinal symptoms; insomnia; hyperactivity	Rind on hard cheeses such as Edam	Cosmetics, pharmaceutical preparations
Regarded as safe in food use; may lower cholesterol levels	Infant formula, ice cream, pickles, icings, toppings, chutney, cheese, confectionery	Cosmetics, animal feed, detergent, adhesives
Regarded as safe in food use at low levels	Processed foods	Animal and poultry feed
Regarded as safe in food use; may have beneficial effects on the body	Nutritional bars, soups, yoghurt, beverages	
Chronic headaches; allergic reactions	Cheese preparation	Cosmetics, tablets, lozenges, eye drops

Names	Number	Function	Code	
Magnesium carbonates (synthetic; from magnesium sulphate and sodium carbonate)	E504	Anticaking agent Alkali	☺	
Magnesium chloride (synthetic; from hydrochloric acid and magnesium oxide/ hydroxide)	E511	Firming agent Buffer	☺	
Magnesium diglutamate (may be GM; contains MSG)	E625	Flavour enhancer	☺	
Magnesium hydroxide (from magnesium chloride and sodium hydroxide or precipitation of seawater)	E528	Acidity regulator	☺☺	
Magnesium oxide (from magnesite ores)	E530	Anticaking agent Firming agent	☺☺	
Magnesium phosphates	E343	Anticaking agent	☺☺	
Magnesium salts of fatty acids (may be of ANIMAL origin)	E470b	Emulsifier Stabiliser	☺☺	

Potential Effects	Possible Food Use	Other Uses
Regarded as safe in food use at low levels; excess may have a laxative effect	Sour cream, ice cream, canned peas, table salt	Baby powder, face powder, perfume carrier
Regarded as safe in food use at low levels; excess may have a laxative effect	Infant formula, salt substitute, non-alcoholic beverages	
Regarded as safe in food use at low levels; contains MSG; laxative effect with excess	Low sodium salt substitute	
Regarded as safe in food use at low levels	Manufacture of some caramels, canned peas, cheese manufacture	Dentifrices, skin cream
Regarded as safe in food use	Frozen dairy products, canned peas	
Regarded as safe in food use	Dried milk, milk powder	Dietary supplement
Regarded as safe in food use	Cake mixes, oven-ready fries	

Names	Number	Function	Code	
Magnesium silicate; magnesium trisilicate	E553a	Anticaking agent	☺	
Malic acid (from fruit or made synthetically)	E296	Acidity regulator Antioxidant	☺	
Maltitol; maltitol syrup (from maltose)	E965	Sweetener Stabiliser	☺	
Mannitol (prepared from seaweed)	E421	Sweetener Humectant	☹	
Metatartaric acid (from tartaric acid)	E353	Acidity regulator Sequestrant	☺☺	
Methylcellulose (made from wood pulp or chemical cotton; may be GM)	E461	Thickener Stabiliser	☺	

FOOD ADDITIVES

	Potential Effects	Possible Food Use	Other Uses
	Regarded as safe in food use at low levels; kidney stones	Vanilla powder	Talcum powder, shampoo, antacid
	Regarded as safe in food use; skin and mucous membrane irritation; may aggravate herpes simplex symptoms	Sweetened coconut, tinned oxtail soup, low calorie soft drinks, wines	Cosmetics, hair lacquer
	Regarded as safe in foods at low levels; laxative effect with excess	Low calorie foods, dried fruit, confectionery	Brewing industry
	Hypersensitivity reactions; nausea; vomiting; diarrhoea; hives; NRC and diabetics; kidney dysfunction; gastric irritation; anaphylaxis	Carbohydrate modified foods or low calorie foods, chewing gum, sweets, jams, jellies	Hand cream, hair grooming products
	Regarded as safe in food use at low levels	Wine, sparkling wine, fruit and vegetable juices	
	Regarded as safe in food use at low levels; laxative effect with excess	Infant formula, ice cream, toppings, pickles, chutney, confectionery, imitation fruit, soup	Cosmetics, hand cream and lotion, sun cream, slimming aids

Names	Number	Function	Code	
Methyl p-hydroxybenzoate (methylparaben; see Parabens in section 2)	E218	Preservative	☹	
Microcrystalline wax (obtained from crude oil)	E905	Glazing agent	☺☺	
Mixed acetic and tartaric acid esters of mono-and diglyc-erides of fatty acids (may be of ANIMAL origin)	E472f	Emulsifier Stabiliser	☺☺	
Mono-and diacetyltartaric acid esters of mono-and diglycerides of fatty acids (may be of ANIMAL origin)	E472e	Emulsifier	☺☺	
Mono-and diglycerides of fatty acids (may be of ANIMAL origin; may be GM)	E471	Emulsifier Stabiliser	☺☺	

FOOD ADDITIVES

	Potential Effects	Possible Food Use	Other Uses
	Asthma; hives; allergic reactions; skin redness, itching and swelling; anaphylaxis	Jellies, preserves, jams, baked goods, fruit juices, salad dressings	Cosmetics, skin and hair care products
	Regarded as safe in food use	Cake decorations	Tablet coatings, laminating paper and foils
	Regarded as safe in food use	Processed bread	Cosmetic cream
	Regarded as safe in food use	Bread, frozen pizza, gravy granules, hot chocolate mix	Cosmetic cream
	Regarded as safe in food use	Cakes, hot chocolate mix, sponge puddings, margarine, ice cream, quick custard mix	

Names	Number	Function	Code	
Monoammonium glutamate (from glutamic acid; contains MSG; may be GM)	E624	Flavour enhancer	☺?	
Monopotassium glutamate (may be GM)	E622	Flavour enhancer	☹	
Monosodium glutamate (MSG; monosodium salt of glutamic acid; may be GM)	E621	Flavour enhancer	☹	
Monostarch phosphate (modified starch; see Starch – Modified in section 2)	E1410	Thickener Stabiliser	☺?	
Montan acid esters (from fossilized vegetable wax)	E912	Surface coating		
Natamycin (pimaricin; fungicide produced from *Streptomyces natalensis*)	E235	Preservative	☹	

Potential Effects	Possible Food Use	Other Uses
Contains MSG; allergic reactions	Low sodium salt substitute	
See Monosodium Glutamate (E621)	Low-sodium salt substitute	
Bronchospasm; heart palpitations; abdominal discomfort; irritability; fibromyalgia; nausea; depression; headache; migraine; asthma; blurred vision; vertigo; sight impairment; teratogenic; aspirin sensitive people may wish to avoid; NRC	May be found in packet soup, quick soup, flavoured noodles, textured protein, malt extract, yeast extract, TVP, soy sauce, gelatine, flavourings (chicken, beef, pork, smoke)	Hidden sources of MSG including soap, cosmetics, shampoo, hair conditioner, most "live virus" vaccines
Uncertainties exist about the safety of modified starches especially in infants	Various foods	
Insufficient information to assess	Protective layer on fruit skins; coating on foods	
Moderately toxic by ingestion; nausea; vomiting; diarrhoea; anorexia; mild skin irritation	Cured, processed meats, cheese rind	Drug used to treat fungal infections of the eyes and eyelids

Names	Number	Function	Code	
Neohesperidine DC (produced from Seville oranges)	E959	Sweetener Flavour enhancer	☺☺	
Nisin (crystals from the bacteria *Streptococcus lactis*)	E234	Preservative Anti-microbial	☺?	
Nitrogen	E941	Propellant Packing gas	☺☺	
Nitrous oxide (laughing gas)	E942	Propellant	☺☺	
Octyl gallate (synthetic; salt of gallic acid)	E311	Antioxidant	☺?	
Orthophenyl phenol (synthetic; from sodium hydroxide and chlorobenzene; see phenol in section 2)	E231	Preservative	☹	
Oxidised polyethylene wax (synthetic; from petroleum)	E914	Humectant	☺?	

Potential Effects	Possible Food Use	Other Uses
Regarded as safe in food use	Cheese, chewing gum, snack foods, instant coffee	Speciality beers
The European Parliament said in 2003 that it should not be used as it could cause antibiotic resistance in humans	Processed cheese, canned vegetables and fruit, semolina and tapioca puddings	
Regarded as safe in food use	Freezing and vacuum packing of foods	Preservative in cosmetics
Regarded as safe in food use	Flour (bleached), food aerosols	Whipped cosmetic cream
Mildly toxic by ingestion; allergic reactions; gastric irritation; aspirin sensitive people may wish to avoid	Edible fats and oils, reduced fat spread, margarine, dripping, salad oil	
Nausea; convulsions; vomiting; circulatory collapse; respiratory failure; cardiac arrest; coma; carcinogenic	Used in the manufacture of many food additives	
Carcinogenic?; kidney and liver damage?	Used as a protective coating on fruits and vegetables	

Names	Number	Function	Code	
Oxidised starch (synthetic; modified starch; see Starch - Modified in section 2)	E1404	Thickener Stabiliser	☺?	
Oxygen	E948	Packing gas	☺☺	
Paprika extract; cap-santhin; capsorubin (extracted from peppers)	E160c	Colouring (orange to red)	☺☺	
Patent Blue V (coal tar or azo dye; banned in some countries)	E131	Colouring (bluish violet)	☹	
Pectins (from apple residue and orange pith)	E440	Thickener Stabiliser	☺	
Phosphated distarch phosphate (modified starch; see Starch – Modified in section 2)	E1413	Thickener Stabiliser	☺?	

FOOD ADDITIVES

Potential Effects	Possible Food Use	Other Uses
Uncertainties exist about the safety of modified starches especially in infants	Casserole mix, batter mix, confectionery	
Regarded as safe in food use	Manufacture of cider	
Regarded as safe in food use; may have beneficial health effects	Cheese slices, chicken pies, condiments, soup	Poultry feed
Asthma; gastrointestinal symptoms; anaphylaxis; hives; hyperactivity; allergic reactions; aspirin sensitive people may wish to avoid	Processed foods, soft drinks	Medical diagnostic dye
May provoke symptoms in those who react to MSG; may have beneficial health effects	Jams, marmalade, ice cream, dessert mix, fruit jelly	Antidiarrheal medicines
Uncertainties exist about the safety of modified starches especially in infants	Flavoured yoghurt, ice cream, canned foods for infants	

Names	Number	Function	Code	
Phosphoric acid (made from phosphate rock)	E338	Acidity regulator	☺?	
Plain caramel (may be from sugar beet, sugar cane or maize starch; may be GM)	E150a	Colouring (dark brown)	☺?	
Polydextrose (similar to cellulose)	E1200	Humectant Stabiliser	☺	
Polyethylene glycol 6000 (see also Polyethylene Glycol in section 2)	No number	Antifoaming agent Solvent	☺?	
Polyglycerol esters of fatty acids (may be of ANIMAL origin; may be GM)	E475	Emulsifier Thickener	☺☺	

Potential Effects	Possible Food Use	Other Uses
Regarded as safe in food use at low levels; excess may lead to tooth erosion and calcium loss in bones	Cheese products, soft drinks, jellies, sweets	
Gastrointestinal problems; needs to be tested for mutagenic, teratogenic, subacute and reproductive effects	Brown bread, cola drinks, chocolate, ice cream, jams, sweets	
Regarded as safe in food use at low levels; excess may have a laxative effect; NRC	Low calorie foods, yoghurt, custard powder, ice cream, confectionery	
Has caused renal failure when used on burn victims; heart problems; mutagenic	Carrier for sweeteners	Antiperspirant, baby products, protective cream, lipstick
Regarded as safe in food use	Mayonnaise, cake mixes, imitation cream, coffee whitener, icings	

Names	Number	Function	Code	
Polyglycerol polyricinoleate (from castor oil and glycerol esters; may be of ANIMAL origin; may be GM)	E476	Emulsifier	☺☺	
Polyoxyethylene (40) stearate (may be of ANIMAL origin)	E431	Emulsifier Defoamer	☺?	
Polyoxyethylene sorbitan monolaurate; polysorbate 20 (synthetic; from sorbitol; may be of ANIMAL origin)	E432	Emulsifier Stabiliser	☹	
Polyoxyethylene sorbitan monooleate; polysorbate 80	E433	Emulsifier Stabiliser	☹	
Polyoxyethylene sorbitan monopalmitate; polysorbate 40 (synthetic; from sorbitol; may be of ANIMAL origin)	E434	Emulsifier	☹	
Polyoxyethylene sorbitan monostearate; polysorbate 60	E435	Emulsifier Stabiliser	☹	

Potential Effects	Possible Food Use	Other Uses
Regarded as safe in food use	Chocolate, drinking chocolate, chocolate ice cream, toppings, icings, cake mixes	
Skin tumours in mice; may facilitate the penetration of cancer-causing additives	Processed foods; frozen desserts	Hand cream and lotion
Polysorbates can contain residues of harmful chemicals; can increase the absorption of fat soluble substances	Bakery products, confectionery, soup, desserts	Cosmetic cream and lotion
Associated with the contaminant 1,4 dioxane known to cause cancer in animals	Icing, frozen custard, sherbet, mayonnaise, ice cream, pickles	Vitamin and mineral supplements
Polysorbates can contain residues of harmful chemicals; can increase the absorption of fat soluble substances	Whipped cream, soup, beverages, confectionery	
Associated with the contaminant 1,4 dioxane known to cause cancer in animals	Cakes, cake mixes, icing, confectionery, beverage mixes	

Names	Number	Function	Code
Polyoxyethylene sorbitan tristearate; polysorbate 65	E436	Emulsifier	☹
Polyphosphates (salts of phosphoric acid)	E452	Emulsifier Stabiliser	☺☺
Polyvinylpolypyrrolidone (synthetic protein)	E1202	Clarifying agent	☹
Polyvinylpyrrolidone (PVP)	E1201	Clarifying agent Stabiliser	☹
Ponceau 4R; Cochineal Red A (brilliant scarlet; monoazo dye; banned in some countries)	E124	Colouring (red)	☹
Potassium acetate (potassium salt of acetic acid)	E261	Acidity regulator Preservative	☺

Potential Effects	Possible Food Use	Other Uses
Polysorbates can contain residues of harmful chemicals; can increase the absorption of fat soluble substances	Ice cream, frozen custard, cake icings and fillings	
Regarded as safe in food use	Meat and fish products, processed cheese, baking powder, cola drinks	Fertilisers, detergents
See Polyvinylpyrrolidone (E1201)	Used to clarify sparkling wine and vinegar	Hairspray
Lung and kidney damage; gas; tumours; allergic contact dermatitis in sunscreen; skin sensitisation; carcinogenic	Beer, wine and vinegar manufacture	Hairspray and lacquers, shampoo, sunscreen
Asthma; hay fever; hives; aspirin sensitive people may wish to avoid; hyperactivity; carcinogenic	Packet trifle mix, jelly crystals, jam, packet cake mix, dessert toppings, tomato soup	
People with impaired kidney function or cardiac disease may wish to avoid	Pickles, salad cream, chutney, cheese, brown sauce, fruit sauce	Diuretic

Names	Number	Function	Code	
Potassium adipate (potassium salt of adipic acid)	E357	Acidity regulator Buffer	☺	
Potassium alginate (potassium salt of alginic acid)	E402	Thickener Stabiliser	☺	
Potassium aluminium silicate	E555	Anticaking agent	☺?	
Potassium benzoate (synthetic; potassium salt of benzoic acid)	E212	Preservative	☹	
Potassium carbonates (inorganic salt of potassium)	E501	Acidity regulator Stabiliser	☺?	
Potassium chloride	E508	Gelling agent	☺?	

	Potential Effects	Possible Food Use	Other Uses
	See Adipic Acid (E355)	Low salt foods	Boiler water
	Alginates may have beneficial effects on health; alginates inhibited absorption of essential nutrients in some animal tests	Ice cream, yoghurt, custard mix	Processed foods, soda water
	See Aluminium (E173)	Cheese and cheese products	
	Asthma; hives; eczema; allergic reactions; gastric irritation; aspirin sensitive people may wish to avoid	Low joule jams and spreads, chilli paste, glace cherries	Cosmetics
	Regarded as safe in food use; dermatitis of the scalp, forehead and hands; eye irritation; upper respiratory tract irritation	Confections, cocoa products, low sodium salt substitute, soda water	Shampoo, permanent wave lotion, washing powder
	Regarded as safe in food use at low levels; intestinal ulcers; NRC; cardiovascular, liver and respiratory toxicity	Reduced sodium foods, low sodium salt substitute	Cosmetics, fertiliser

Names	Number	Function	Code	
Potassium citrates (potassium salt of citric acid)	E332	Acidity regulator Stabiliser	☺	
Potassium ferrocyanide (by-product in purification of coal gas; banned in the USA)	E536	Anticaking agent Colouring	☺?	
Potassium gluconate (potassium salt of gluconic acid)	E577	Sequestrant Buffer	☺☺	
Potassium hydrogen sulphite (see Sulphites in section 2)	E228	Preservative	☹	
Potassium hydroxide	E525	Alkali Emulsifier	☺?	
Potassium lactate (potassium salt of lactic acid; may be of ANIMAL origin; may be GM)	E326	Acidity regulator Humectant	☺	

Potential Effects	Possible Food Use	Other Uses
May provoke symptoms in those who react to MSG; may interfere with the results of laboratory tests for pancreatic, blood and liver function	Confectionery, jellies, preserves	
Kidney toxicity; skin irritation; may be harmful by inhalation, ingestion and skin absorption	Salts and condiments	
Regarded as safe in food use	Processed foods, Soda water	
Asthma; skin reactions; anaphylaxis; gastric irritation; hyperactivity	Low joule jam, dried fruits	
Regarded as safe in food use; skin irritation and nail damage in cuticle removers; tumours in mice when applied to the skin	Extracting colour from annatto seed, cacao products	Hand lotion, shaving cream, cuticle removers
Regarded as safe in food use; people with lactose intolerance may wish to avoid	Biscuits, cheese, confectionery, foods for infants	

Names	Number	Function	Code	
Potassium malate (potassium salt of malic acid)	E351	Acidity regulator	☺☺	
Potassium metabisulphite (see Sulphites in section 2)	E224	Preservative Antioxidant	☹	
Potassium nitrate (banned in some countries; see Nitrates in section 2)	E252	Preservative Colour fixative	☹	
Potassium nitrite (banned in some countries; see Nitrites in section 2)	E249	Preservative Colour fixative	☹	
Potassium phosphates	E340	Acidity regulator Stabiliser	☺☺	

	Potential Effects	Possible Food Use	Other Uses
	Regarded as safe in food use	Fruit drinks, soft drinks, sweetened coconut	
	Asthma; hives; behavioural problems; gastric irritation; anaphylaxis	Cheese and cheese products, home wine brewing kits	Bleaching straw
	Asthma; kidney inflammation; behavioural problems; dizziness; headache; may form nitrosamines; not permitted in foods for babies under 6 months	Prosciutto ham, cured, processed meats	Tobacco, matches
	Asthma; kidney inflammation; behavioural problems; dizziness; headache; may form nitrosamines; not permitted in foods for babies under 6 months; inhibits oxygen in the blood; carcinogenic	Corned, cured, pickled, manufactured and pressed meats	
	Regarded as safe in food use	Frozen egg products, production of champagne	Shampoo, cuticle remover

Names	Number	Function	Code	
Potassium propionate (potassium salt of propionic acid)	E283	Preservative	☺?	
Potassium sorbate (potassium salt of sorbic acid)	E202	Preservative	☺?	
Potassium sulphates (naturally occurring mineral)	E515	Acidity regulator	☺	
Potassium tartrates (potassium salt of tartaric acid)	E336	Acidity regulator Stabiliser	☺?	
Processed eucheuma seaweed	E407a	Thickener Stabiliser	☺?	

Potential Effects	Possible Food Use	Other Uses
Asthma; learning difficulties; headache; behavioural problems; migraine; see also Calcium Propionate (E282)	Bread, biscuits, cakes, pastries and other flour products	Cosmetics
Allergic reactions; asthma; skin irritation; behavioural problems	Bread, cheese, baked goods, cheesecake, wine making, chocolate	Cosmetics, cigarettes
Regarded as safe in food use at low levels; large doses cause severe gastrointestinal bleeding	Low sodium salt substitute, brewing industry	Cosmetics, medicines
Should be avoided by people with impaired kidney or liver function; high blood pressure; oedema or cardiac failure	See Tartaric acid (E334)	
Uncertainties exist as to the safety of this additive	Processed foods	

Names	Number	Function	Code	
Propane-1,2-diol; propylene glycol (made synthetically from propylene or glycerol or propylene oxide)	E1520	Humectant Solvent	☹	
Propane	E944	Propellant Aerator	☺?	
Propane-1,2-diol alginate (propylene glycol ester of alginic acid)	E405	Thickener Stabiliser	☺?	
Propane-1,2-diol esters of fatty acids (may be of ANIMAL origin; may be GM)	E477	Emulsifier Aerating agent	☺☺	
Propionic acid (obtained from wood pulp, waste liquor and by fermentation)	E280	Preservative	☺?	

Potential Effects	Possible Food Use	Other Uses
Contact dermatitis; lactic acidosis; dry skin; respiratory, immuno- and neurotoxicity; CNS depression and kidney damage in animals	Confectionery, baked goods, chocolate products, sweetened coconut, toppings	Suntan lotion, toothpaste, lipstick, baby lotion, pesticides, antifreeze
May be narcotic in high concentrations; neurotoxicity; being reassessed for safety	Foamed and sprayed foods	Cosmetics in aerosols
Can cause allergic reactions; reduced growth and loose stools in animal studies; alginates inhibited absorption of essential nutrients in some animal tests	Frozen custard, salad dressing, ice cream, fruit sherbet	Cosmetics
Regarded as safe in food use	Whipped toppings, cakes	
Migraines; skin irritation; headaches; harmful to aquatic organisms	Ice cream, baked goods, sweets, processed cheeses	Bakery use

Names	Number	Function	Code	
Propyl gallate (synthetic ester of gallic acid)	E310	Antioxidant	☺?	
Propyl p-hydroxybenzoate (propylparaben; see Parabens in section 2)	E216	Preservative	☹	
Pullulan (a glucan produced from corn starch by a fungus; may be GM)	E1204	Glazing agent Film former	☺	
Quillaia extract (quillaja extract; banned in some countries)	E999	Foaming agent Flavouring	☺	

Potential Effects	Possible Food Use	Other Uses
Asthma; contact dermatitis; gastric irritation; aspirin sensitive people may wish to avoid; not permitted in foods for babies and young children	Dairy blend, edible fats and oils, reduced fat spread, chewing gum, margarine, salad oil, peanut butter	Cosmetic cream and lotion
Asthma; hives; allergic reactions; skin redness, itching and swelling; anaphylaxis	Beverages, baked goods, sweets, jellies, preserves, fruit flavourings	Cosmetics; shampoo, foundation cream
Regarded as safe in food use; excess may cause bloating & mild gastrointestinal upset	Food supplements in capsule & tablet form, jams, jellies, confectionery	Mouthwash, breath fresheners
Regarded as safe in food use; gastrointestinal irritation; large doses can cause liver damage, respiratory failure, convulsions and coma	Soft drinks, ice cream, sweets	

Names	Number	Function	Code	
Quinoline Yellow (synthetic; azo dye; banned in some countries; may be of ANIMAL origin)	E104	Colouring (dull yellow to greenish yellow)	☹	
Red 2G (synthetic; azo dye; banned in many countries)	E128	Colouring (red)	☹	
Riboflavin; riboflavin-5'-phosphate (vitamin B2; may be GM)	E101	Colouring (yellow or orange)	☺☺	
Saccharin and its Na, K and Ca salts (prepared from toluene; banned or restricted in many countries)	E954	Artificial sweetener	☹☹	
Salt of aspartame-acesulfame	E962	Artificial sweetener	☹	
Shellac (from resin produced by the Lac insect)	E904	Glazing agent	☺	

	Potential Effects	Possible Food Use	Other Uses
	Asthma; hives; skin rash; hyperactivity; anaphylaxis; aspirin sensitive people may wish to avoid; carcinogenic	Beverages, processed foods	Cosmetic dye, lipstick, soap, toothpaste, hair products, cologne
	Asthma; gastrointestinal symptoms; hyperactivity; angioedema; chronic hives; aspirin sensitive people may wish to avoid; carcinogenic	Cooked meat products, burger meats, sausages	
	Regarded as safe in food use; has beneficial effects in the body	Baby cereals, enriched breads, peanut butter, breakfast cereals	Vitamin tablets
	Hives; pruritis; NRC; eczema; nausea; diarrhoea; diuresis; headache; mutagenic; carcinogenic; teratogenic	Diet soft drinks, sugar substitute	Chewable aspirin, pharmaceutical preparations
	See Aspartame (E951) and Acesulfame K (E950)	Instant pudding mix, dairy shake mixes, chewing gum, chocolate	
	Regarded as safe in food use; allergic contact dermatitis; skin irritation	Sweets, waxed fruit	Cosmetics, hair lacquer, tablets, mascara, jewellery

Names	Number	Function	Code	
Silicon dioxide (salts from silicic acid)	E551	Anticaking agent	☺☺	
Silver (a naturally occurring metal)	E174	Colouring (metallic)	☺?	
Sodium acetate (sodium salt of acetic acid)	E262	Acidity regulator Preservative	☺	
Sodium adipate (sodium salt of adipic acid)	E356	Acidity regulator Firming agent	☺☺	
Sodium alginate (sodium salt of alginic acid)	E401	Thickener Stabiliser	☺	

	Potential Effects	Possible Food Use	Other Uses
	Regarded as safe in food use; has beneficial health effects	Dried egg products, beverage whitener, salt substitutes	Beer production, animal feed, paper and paperboard
	Toxic in very large doses; should not be consumed; accumulates in tissues; argyria (blue-grey skin); kidney damage	External decoration on cakes, "silver dragees", silver-coloured almonds	Cosmetics, nail polish
	Regarded as safe in food use; skin and eye irritation; moderate toxicity by ingestion	Sweets, jams, jellies, soup mixes, snack foods, cereals	Cosmetics, textiles, photographic and dye processes
	Regarded as safe in food use	Beverages, baked goods, processed cheese, snack foods	
	Alginates may have beneficial health effects; alginates inhibited absorption of essential nutrients in animal tests	Frozen desserts, jams, fruit jelly preserves	Baby lotion, wave sets, shaving cream

Names	Number	Function	Code	
Sodium aluminium phosphate	E541	Emulsifier Acidity regulator	😐?	
Sodium aluminium silicate	E554	Anticaking agent	😐?	
Sodium ascorbate (synthetic; sodium salt of ascorbic acid)	E301	Antioxidant	😊😊	
Sodium benzoate (sodium salt of benzoic acid)	E211	Preservative	😞	
Sodium carbonates (mostly manufactured synthetically)	E500	Acidity regulator Anticaking agent	😐?	

	Potential Effects	Possible Food Use	Other Uses
	Regarded as safe in food use at low levels; people with kidney or heart disease may wish to avoid or limit intake; see Aluminium (E173)	Self-raising flour, various cheeses	
	Regarded as safe in food use at low levels; see Aluminium (E173)	Beverage whitener, baking powder, dry soup mix	Barrier cream, depilatories
	Regarded as safe in food use	Foods for infants, frozen fish, wine, vinegar, beer	Cosmetics
	Asthma; hives; contact dermatitis; hay fever; mouth and skin irritation; hyperactivity; anaphylaxis; aspirin sensitive people may wish to avoid	Bottled soft drinks, fruit juice, jams, pickles, condiments, baked goods, tomato paste	Toothpaste, eye cream, medical diagnostic aid for liver function
	Regarded as safe in food use at low levels; contact can cause forehead, scalp and hand rash; respiratory distress; large doses can cause gastrointestinal bleeding, vomiting, diarrhoea, shock and death	Confectionery, custard mix, flour products, ice cream mix, soup, jams, chocolate, malted milk powder	Mouthwash, soap, bath salts, vaginal douches, shampoo, laundry detergent, paper and glass manufacture

Names	Number	Function	Code	
Sodium citrates (sodium salt of citric acid)	E331	Acidity regulator Emulsifier	☺	
Sodium erythorbate (sodium salt of erythorbic acid)	E316	Antioxidant	☺	
Sodium ethyl p-hydroxybenzoate (synthetic; from benzoic acid; see Parabens in section 2)	E215	Preservative	☹	
Sodium ferrocyanide (synthetic; from the reaction of cyanide with iron sulphate)	E535	Anticaking agent	☺☺	
Sodium gluconate (sodium salt of gluconic acid)	E576	Acidity regulator Sequestrant	☺	
Sodium hydrogen sulphite (synthetic; see Sulphites in section 2)	E222	Preservative	☹	

FOOD ADDITIVES

	Potential Effects	Possible Food Use	Other Uses
	Can provoke symptoms in those who react to MSG; may alter urinary excretion of some drugs making them either less effective or more potent	Infant formula, cottage cheese, ice cream, evaporated milk, jams, preserves, fruit jellies	Cosmetics
	Regarded as safe in food use; has not been studied for mutagenic and teratogenic effects	Cider, frozen fruit, meat products, baked goods, beverages	Cosmetics, water softener, detergent
	Asthma; numbness in the mouth; hives; gastric irritation; allergic contact dermatitis; aspirin sensitive people may wish to avoid	Dried meat products, snack foods, confectionery	
	Regarded as safe in food use	Salts and condiments	Processing wine
	Regarded as safe in food use; people with heart disease or high blood pressure may wish to avoid or limit	Processed cheese, confectionery, margarine	Metal cleaner, paint stripper, metal plating, rust remover
	Bronchial asthma; chronic hives; skin irritation; mutagenic	Tomato paste, dried fruit, jams, jellies, fruit juices	Mouthwash, hair dye, wart remover

Names	Number	Function	Code	
Sodium hydroxide (made by electrolysis of sodium chloride brine)	E524	Emulsifier	☺	
Sodium lactate (sodium salt of lactic acid)	E325	Humectant Bulking agent	☺	
Sodium malates (sodium salt of malic acid)	E350	Acidity regulator Humectant	☺☺	
Sodium metabisulphite (synthetic; sodium salt of sulphurous acid; see Sulphites in section 2)	E223	Preservative	☹	
Sodium methyl p-hydroxybenzoate (see Parabens in section 2)	E219	Preservative	☹	
Sodium nitrate (sodium salt of nitric acid; banned in some countries; see Nitrates in section 2)	E251	Preservative	☹	

Potential Effects	Possible Food Use	Other Uses
Regarded as safe in food use; contact can cause dermatitis; concentrate is hazardous	Modifier for starch, glazing of pretzels	Shampoo, shaving cream, liquid drain cleaner
Regarded as safe in food use; people with lactose intolerance may wish to avoid	Biscuits, uncured hams	Moisturiser, skin and hair products
Regarded as safe in food use	Fruit drink, soft drinks, sweetened coconut	Anti-aging products
Asthma (life threatening attacks); hay fever; chronic hives; atopic dermatitis; harmful to aquatic organisms	Bread and flour products, jellies, dried fruits, tomato paste, maraschino cherries	Hair products, bath preparations, underarm deodorant
Asthma; hives; allergic reactions; skin redness, itching and swelling; anaphylaxis	Dried meat products, snack foods, confectionery	
Nausea; vomiting; dizziness; headaches; migraine; may affect thyroid gland; not permitted in foods for babies less than 6 months	Prosciutto ham, manufactured meats	

Names	Number	Function	Code	
Sodium nitrite (synthetic; banned in some countries; see Nitrites in section 2)	E250	Preservative	☹	
Sodium orthophenyl phenol (see Phenol in section 2)	E232	Preservative Antifungal	☹	
Sodium phosphates (sodium salts of phosphoric acid)	E339	Acidity regulator Emulsifier	☺	
Sodium, potassium and calcium salts of fatty acids (may be of ANIMAL origin)	E470a	Emulsifier Stabiliser	☺☺	
Sodium potassium tartrate (sodium and potassium salt of tartaric acid; Rochelle salt)	E337	Acidity regulator Stabiliser	☺	

FOOD ADDITIVES

	Potential Effects	Possible Food Use	Other Uses
	Nausea; headaches; dizziness; not permitted in foods for babies less than 6 months; toxic to aquatic organisms	Canned, cured, manufactured and pressed meats, sausages, bacon	Anticorrosive in some cosmetics
	Vomiting; convulsions; irritation of nose and eyes; depigmentation; photosensitiser	Sprayed onto fruit skins, products containing fruit skins, marmalade	Cosmetics, detergents, cooling fluids
	Regarded as safe in foods at low levels; contact can cause skin irritation; erythema; blisters	Frozen desserts, noodle and macaroni products, cheese spread	Manufacture of nail polish and detergents
	Regarded as safe in food use	Cake mixes, oven ready fries	
	Regarded as safe in food use; people with oedema, high blood pressure, cardiac failure, kidney or liver damage advised to avoid	Confectionery, jams, fruit jelly preserves, cheese, manufacture of baking powder	Silvering of mirrors, mouthwash, cathartic in medicinal use

Names	Number	Function	Code	
Sodium propionate (sodium salt of propionic acid)	E281	Preservative	🙂?	
Sodium propyl p-hydroxybenzoate (see Parabens in section 2)	E217	Preservative	☹	
Sodium stearoyl 2-lactylate (from lactic acid and fatty acids; may be of ANIMAL origin; may be GM)	E481	Emulsifier Stabiliser	🙂🙂	
Sodium sulphates (sodium salts of sulphuric acid)	E514	Acidity regulator Preservative	🙂?	
Sodium sulphite (synthetic; sodium salts of sulphurous acid; see Sulphites in section 2)	E221	Preservative	☹	

Potential Effects	Possible Food Use	Other Uses
Behavioural problems; skin irritation; learning difficulties; asthma; gastric irritation; headache; migraines; see also Calcium Propionate (E282)	Confectionery, baked goods, frostings, cakes	Cosmetics, medical treatment of fungal infections of the skin
Asthma; hives; numbness in the mouth; gastric irritation; allergic contact dermatitis; aspirin sensitive people may wish to avoid	Dried meat products, snack foods, confectionery	
Regarded as safe in food use at low levels	Biscuits, bread, cakes, cake icings, fillings and toppings	
Regarded as safe in foods at low levels; skin irritation; gastrointestinal irritation; people with poor kidney or liver function should avoid	Chewing gum base, biscuits, tuna fish	Manufacture of dye, soap, detergents, glass and paper
Asthma; gastric irritation; skin rash; nausea; diarrhoea; destroys vitamin B content in food; those with poor liver or kidney function should avoid	Cut fruits, dried fruit, maraschino cherries, prepared fruit pie mix, frozen apples	

Names	Number	Function	Code	
Sodium tartrates (sodium salts of tartaric acid)	E335	Acidity regulator Sequestrant	☺☺	
Sodium tetraborate; borax (banned or replaced in most countries)	E285	Preservative	☹	
Sorbic acid (may be from the berries of Mountain Ash or synthetic from chemicals)	E200	Preservative Humectant	☺?	
Sorbitan monolaurate (synthetic; from sorbitol and lauric acid)	E493	Emulsifier Stabiliser	☺☺	
Sorbitan monooleate (from sorbitol and oleic acid; may be of ANIMAL origin)	E494	Emulsifier	☺	
Sorbitan monopalmitate	E495	Emulsifier Stabiliser	☺	

Potential Effects	Possible Food Use	Other Uses
Regarded as safe in food use	Cheese, artificially sweetened jelly, meat products	
Toxic; rarely used with foods; chronic exposure can cause red peeling skin, seizures and kidney failure	Meats imported from the USA?	Shaving cream, cold cream, foundation cream, insecticides
Allergic reactions; asthma; contact dermatitis; erythema; skin irritation; behavioural problems	Frozen pizza, pie fillings, cheese, cheesecake, cheese spread, chocolate syrup	Cosmetics, mouthwash, toothpaste, ointments, dental cream
Regarded as safe in food use	Tea concentrates including fruit and herbal	
Regarded as safe in food use; contact dermatitis; allergic reactions	Tea concentrates including fruit and herbal	Pharmaceuticals
Regarded as safe in food use	Cake mixes	

Names	Number	Function	Code	
Sorbitan monostearate (synthesised from sorbitol and stearic acid; may be of ANIMAL origin; may be GM)	E491	Emulsifier Glazing agent	☺	
Sorbitan tristearate (prepared from sorbitol and stearic acid; may be of ANIMAL origin; may be GM)	E492	Emulsifier	☺	
Sorbitol; sorbitol syrup (may be synthesised from glucose)	E420	Humectant Sweetener	☺?	
Soybean hemicellulose (extracted from soybean fibre)	E426	Emulsifier Thickener Stabiliser	☺?	

Potential Effects	Possible Food Use	Other Uses
Regarded as safe in food use at low levels; high dietary levels can cause intralobular fibrosis; growth retardation; liver enlargement	Confectionery, ice cream, flavoured milk, bakery wares, cake mix, icing	Cosmetic cream and lotion, suntan cream, skin cream, deodorant
Regarded as safe in food use at low levels; high dietary levels can cause intralobular fibrosis; growth retardation; liver enlargement	Compounded chocolate, oil toppings, cake mixes	Insecticides, nail strengthening cream
Excess intake can cause intestinal cramps; diarrhoea; gastrointestinal disturbance; bloating; cataracts; may alter absorption of drugs so they are either more toxic or less effective; on NIH hazards list	Confectionery, dried fruit, chewing gum, chocolate, lollies	Cosmetics, hair spray, shampoo, mouthwash, toothpaste, embalming fluid, antifreeze
May cause adverse reactions in those allergic to soybean	Baked goods, jelly confectionery, rice, noodles	

Names	Number	Function	Code	
Stannous chloride (salt of tin)	E512	Antioxidant Colour retention agent	😐?	
Starch aluminium octenyl succinate (synthetic; may be GM; see Starch – Modified in section 2)	E1452	Thickener	🙁	
Starch sodium octenyl succinate (modified starch; see Starch – Modified in section 2)	E1450	Thickener Stabiliser	😐?	
Stearyl tartrate (from tartaric acid; may be of ANIMAL origin; banned in some countries)	E483	Emulsifier Stabiliser	😐?	
Succinic acid (prepared from acetic acid)	E363	Flavour enhancer Acidity regulator	🙂	

Potential Effects	Possible Food Use	Other Uses
Low systemic toxicity, but may be irritating to skin and mucous membranes; harmful to aquatic organisms	Canned asparagus, canned fruit juice, canned soft drinks	Manufacture of dye
Uncertainties exist about the safety of modified starches especially in infants; respiratory toxicity; may contain harmful impurities; see also Aluminium (E173)	Encapsulated vitamin preparations in food supplements	Cosmetics, sunscreen, tanning oil, aftershave, shampoo
Uncertainties exist about the safety of modified starches especially in infants	Essence, salad dressing, beverage whitener	
Regarded as safe in food use; concerns exist that it may be carcinogenic	Dough	
Regarded as safe in food use at low levels; excess can cause vomiting and diarrhoea	Salt substitute	Cosmetics, inks, perfume, lacquers, paint, mouthwash

Names	Number	Function	Code	
Sucralose (synthetically prepared from sugar and chlorine)	E955	Artificial Sweetener	😐?	
Sucroglycerides (sucrose ester of fatty acid; may be of ANIMAL origin)	E474	Emulsifier Stabiliser	🙂🙂	
Sucrose acetate isobutrate	E444	Emulsifier Stabiliser	🙂	
Sucrose esters of fatty acids (from sucrose and fatty acids; may be of ANIMAL origin; may be GM)	E473	Emulsifier Stabiliser	🙂	
Sulphite ammonia caramel (synthetic; may be from sugar beet, sugar cane or maize starch, ammonia and sulphite compounds; may be GM)	E150d	Colouring	😐?	

Potential Effects	Possible Food Use	Other Uses
Thymus shrinkage; kidney and liver enlargement in animal studies	Confectionery, fruit spreads, desserts, baked goods	
Regarded as safe in food use	Chocolate milk, cocoa, eggnog, chewing gum, drinking yoghurt	
Regarded as safe in foods; produced liver damage in dogs but not in other species	Citrus flavoured beverages	
Regarded as safe in foods at low levels; large doses can cause nausea, diarrhoea, gas, bloating, abdominal pain; can facilitate uptake of food allergens	Margarine, dairy desserts, chewing gum, chocolate, mayonnaise	
Hyperactivity; soft to liquid stools & increased bowel movements; blood toxicity in rats; inhibits metabolism of B6 in rabbits; serious doubts exist on safety	Stout, cola drinks, chocolate, gravy browning, jams, confectionery	

Names	Number	Function	Code	
Sulphur dioxide (produced by burning sulphur)	E220	Preservative	☹	
Sulphuric acid	E513	Acidity regulator	☺☺	
Sunset Yellow FCF; Orange Yellow S (FD & C Yellow No. 6; synthetic; azo dye; banned in some countries)	E110	Colouring (orange/yellow)	☹☹	
Talc (magnesium silicate)	E553b	Anticaking agent	☹	
Tara gum (derived from a tree native to Peru)	E417	Thickener Stabiliser	☺☺	
Tartaric acid (L-(+)) (by-product of the wine industry)	E334	Antioxidant Food acid	☺	

	Potential Effects	Possible Food Use	Other Uses
	Asthma; broncho spasm; bronchoconstriction; hypotension; anaphylaxis; bronchitis; destroys vitamins A and B1 in food; animal mutagen	Dried fruit, beer, cider, fruit juice, gelatine, wines, pickles, soft drinks, desiccated coconut, vinegar	
	Regarded as safe in food use	Used to modify starch	Brewing industry, cosmetic products
	Asthma; hives, hay fever; abdominal pain; eczema; hives; hyperactivity; aspirin sensitive people may wish to avoid; carcinogenic	Fruit juice cordial, marzipan, packet soup, cereal, confectionery, dry drink powder, canned fish	Cosmetics, hair rinses
	Cancers (stomach and ovarian); cough; vomiting; respiratory problems; tumours; stomach problems	Chocolate, chewing gum base, condiments, confectionery, polished rice	Eye make-up, bath powder, baby powder, animal feed, vitamin supplements
	Regarded as safe in food use	Ice cream, cheese, bakery products, sauces	
	Regarded as safe in food use at low levels; laxative effect from excess	Confectionery, jam, fruit jelly, fruit drink, baking powder, fruit juice, dried egg whites	Denture powder, hair rinses, nail bleaches, depilatories

Names	Number	Function	Code	
Tartaric acid esters of mono- and diglycerides of fatty acids (may be of ANIMAL origin)	E472d	Emulsifier Stabiliser	☺	
Tartrazine (FD&C yellow No. 5; coal tar dye; banned in some countries)	E102	Colouring (lemon-yellow to orange)	☹☹	
Tertiary butyl hydroquinone (TBHQ; contains a petroleum derivative; often used with BHA & BHT; banned in some countries)	E319	Antioxidant	☹☹	
Thaumatin (extracted from the fruit of a West African plant)	E957	Flavour enhancer Sweetener	☺☺	

	Potential Effects	Possible Food Use	Other Uses
	Regarded as safe in food use; headaches	Confectionery, ice cream, bread, dessert toppings, custard mix, cheesecake mix	
	Asthma; hives; dermatitis; headache; hay fever; concentration difficulties; depression; skin rash; learning difficulties; behavioural problems; swelling of lips and tongue; hyperactivity; aggressive behaviour; insomnia; confusion; anaphylaxis; aspirin sensitive people may wish to avoid; NRC; carcinogenic	Confectionery, sweet corn, cheese crackers, soft drinks, mint sauce, mint jelly, fruit juice cordial, canned peas, marzipan, pickles, brown sauce, packet dessert topping, jams, cereal, packaged soups	Cosmetics, wool and silk dye
	Moderately toxic by ingestion; birth defects; tinnitus; allergic contact dermatitis; may be carcinogenic and mutagenic	Edible oils and oil emulsions, muesli and muesli bars, breakfast cereals, lard, fish oil	Sunscreen, body oil, hair colouring, antiperspirant
	Regarded as safe in food use	Chewing gum, Japanese seasonings?	

Names	Number	Function	Code	
Thermally oxidised soya bean oil inter-acted with mono and diglycerides of fatty acids (made from soya; may be GM)	E479b	Emulsifier	☺	
Titanium dioxide (occurs naturally; may contain nanoparticles)	E171	Colouring (white) Opacifier	☺?	
Tocopherols (vitamin E; obtained from edible vegetable oils; may be GM)	E306	Antioxidant	☺☺	
Tragacanth (derived from the plant *Astragalus gummifer*)	E413	Thickener Emulsifier	☺?	
Triammonium citrate (triammonium salt of citric acid)	E380	Acidity regulator	☺	

	Potential Effects	Possible Food Use	Other Uses
	Regarded as safe in food use; topically soybean oil can cause hair damage, allergic reactions and acne-like pimples	Margarine, fat emulsions for frying	
	Regarded as safe in food use; skin contact can cause irritation; limited evidence of cancer in animal studies; see Nanoparticles in section 2	Pan sugar coated confectionery, sweets, chewing gum, icing sugar, jam, jellies	Bath powder, face powder, ointment, sunscreen, marker ink, paints
	Regarded as safe in food use; may be destroyed by freezing	Dairy blend, margarine, salad oil, reduced fat spread	Deodorant, baby preparations, supplements
	Regarded as safe in food use at low levels; adverse reactions such as asthma, abdominal pain, contact dermatitis, dyspnoea, anaphylaxis and constipation can occur but are rare	Sauces, fruit jelly, salad dressing, confections, icings	Shaving cream, rouge, toothpaste, foundation
	May provoke symptoms in those who react to MSG	Wide variety of foods including processed cheese, cheese spread	

Names	Number	Function	Code	
Triethyl citrate (citric acid and ethyl alcohol)	E1505	Thickener Sequestrant	☺	
Triphosphates (made synthetically from phosphate rock)	E451	Emulsifier Stabiliser	☺	
Vegetable carbon (usually from burnt vegetable matter but may be of ANIMAL origin; may be GM; banned in some countries)	E153	Colouring (black)	☺?	
Xanthan gum (may be GM)	E415	Thickener Emulsifier	☺☺	
Xylitol (formally from Birch wood, now made from waste products from the pulp industry)	E967	Humectant Stabiliser	☺	
Zinc acetate (zinc salt of acetic acid)	E650	Flavour enhancer	☺	

FOOD ADDITIVES

Potential Effects	Possible Food Use	Other Uses
Can provoke symptoms in those who react to MSG; citrates may interfere with the results of laboratory tests for blood, liver and pancreatic function	Processed egg whites, smoke flavour	Nail polish, perfume base
Regarded as safe in food use	Processed cheese, meat and fish products, baking powder, cola drinks	
Mildly toxic by ingestion, skin contact and inhalation; may be carcinogenic	Concentrated fruit juice, jams, jellybeans, liquorice, confectionery	Cosmetics
Regarded as safe in food use	Jellies, sweets, dairy products, breakfast cereal, salad dressing	Animal feeds, cigarettes
Regarded as safe in food use at low levels; reported to have beneficial effects on health; large doses cause diarrhoea and flatulence; has caused tumours in rats	Ice cream, chocolate, jams, confectionery, chewing gum, toffee, mints	Toothpaste
Regarded as safe in food use at low levels; large doses can cause nausea and vomiting	Chewing gum	Dietary supplement, animal feed

Section 2

Benzyl salicylate
(synthetic)

UV a...

Betaglucans
(found in oat fibre
and barley)

Thickener

Beta-naphthol
(from naphthalene
from coal tar)

Solvent

COSMETIC INGREDIENTS

Names	Function	Code	
Abietic acid (abietol; from pine rosin)	Stabiliser Texturiser	🙂?	
Acetal (derived from acetaldehyde)	Flavouring Solvent	☹️	
Acetaldehyde (ethanal; may be of ANIMAL origin)	Solvent Intermediate	☹️☹️	
Acetamide MEA (n-acetyl ethanolamine)	Antistatic agent Humectant	🙂?	
Acetaminopropyl trimonium chloride	Antistatic agent	🙂?	
Acetarsol (acetarsone)	Anti-microbial	🙂?	

Potential Effects	Cosmetic Uses	Other Uses
Can cause allergic reactions; skin and mucous membrane irritation; harmful to marine life	Soap manufacture, foaming face wash	Making vinyls, lacquers and plastics
CNS depressant; respiratory depression; cardiovascular collapse; no known skin toxicity; possible high blood pressure; on NIH hazards list	Synthetic perfume	Fruit flavouring in foods, hypnotic in medicine
Mucous membrane irritation; liver damage; kidney, respiratory and neurotoxicity; CNS depression; skin irritation; teratogenic; carcinogenic; harmful to aquatic organisms	Fragrance in cosmetics, perfume manufacture	Silvering of mirrors, synthetic rubber
Mild skin irritation; caused liver cancer in rats; may contain DEA; see Diethanolamine	Hair shampoo and conditioner, skin cream, hair tonic	
See Quaternary Ammonium Compounds	Shampoo, bath soap, conditioner	Detergents
Sensitisation; allergic reactions; lethal dose in mice is only 0.004g/kg of body wt	Mouthwash, toothpaste, feminine hygiene products	

Names	Function	Code	
Acetic acid (occurs naturally in some fruits and plants)	Solvent Rubefacient	😐?	
Acetone (derived by oxidation or fermentation)	Solvent Denaturant	🙁🙁	
Acetonitrile (methylacyanide; precursor of cyanide; on Canadian Hotlist)	Solvent	🙁🙁	
Acetyl tyrosine (may be of ANIMAL origin)	Biological Additive	🙂🙂	
Acetylated lanolin (of ANIMAL origin; may be contaminated with pesticide residues)	Emulsifier Emollient	😐?	
Acetylated lanolin alcohol (see Acetylated lanolin)	Emulsifier Emollient	🙂	

Potential Effects	Cosmetic Uses	Other Uses
Skin irritation; hives; skin rash; caused cancer in rats and mice, orally and by injection; harmful to aquatic organisms	Hand lotion, hair dye, freckle bleaching cream	
Brittle nails; peeling and splitting nails; lung irritation; skin rashes; eye irritation; cardio-vascular, liver and neurotoxicity	Nail polish, nail polish remover	Solvent for airplane glues, cellulose glues, paint thinners
Nervous system poison; skin irritation; gastrointestinal and liver toxicity; teratogenic; fatal if swallowed	Artificial nails remover	Extraction processes
Non-essential amino acid; generally recognised as safe	Cosmetics, suntan cream and liquids	Dietary supplement
Undergoing review for safety; see Lanolin	Baby products, lipstick, eye make-up, cosmetic cream, hair conditioner	
Claimed to be hypoallergenic; may be drying to the skin	Eye make-up, skin moisturiser, bath soap, cologne	

Names	Function	Code	
Acetylmethionyl methylsilanol elastinate (of ANIMAL origin)	Antistatic agent	☺☺	
Acid colours e.g. acid red 14 (black, blue, brown, green, orange, red, violet , yellow; synthetic coal tar/azo dye)	Colourant	☹	
Acrylamide copolymer (acrylamide is derived from acrylonitrile and sulphuric acid)	Film former Thickener	☹	
Acrylates copolomer (synthetic, from petroleum)	Binder Film former	☺?	
Alcohol (ethanol)	Solvent	☺?	
Alkyl benzene sulfonate	Detergent	☺?	

Potential Effects	Cosmetic Uses	Other Uses
Considered safe in cosmetic use	Hair conditioner, skin conditioner	
Many can cause skin, eye and mucous membrane irritation; see Azo Dye and Coal Tar	Tints and dye for hair colouring	
Acrylamide causes liver, reproductive and neurotoxicity; toxic by skin absorption; hazardous to the environment; especially harmful to fish	Nail polish, cosmetics	
Acrylates are strong irritants	Nail polish, blusher, hairspray, mascara	
Implicated in mouth, throat and tongue cancers; contact dermatitis; drying to the skin and hair if used in excess	Mouthwash, facial cleanser, perfume, aftershave	Alcoholic beverages
Believed to be non-toxic orally; drying to the skin; may cause skin irritation	Shampoo, bubble bath	

Names	Function	Code	
Ahnfeltia concinna (derived from algae)	Botanical additive	☺☺	
Allantoin (can be extracted from uric acid, found in comfrey root; may be of ANIMAL origin)	Anti-microbial Oral care agent	☺☺	
Aloe vera, aloe vera gel and aloe vera extract	Botanical additive	☺☺	
Alpha hydroxy acids (AHA's; glycolic acid, lactic acid, tartaric acid, malic acid, citric acid, salicyclic acid, L-alpha hydroxy acid, mixed fruit acids and others; on Canadian Hotlist)	Exfoliant	☹	
Aluminium acetate (mixture including acetic acid and boric acid)	Anti-microbial	☹	

Potential Effects	Cosmetic Uses	Other Uses
No known adverse effects	Skin conditioner, beauty aids	
May accelerate cell growth promoting healing of fractures, scars, wounds; may alleviate psoriasis	Cold cream, hand lotion, hair lotion, after-shave lotion, hair conditioner	
No known adverse effects; reputed to have beneficial and healing effects on the body	Skin cream, deodorant, soap, shaving cream	
Long-term skin damage; swelling, skin discolouration; especially around the eyes; skin blistering; itchiness; rashes; liver toxicity; higher risk of skin cancer, photosensitivity; Do not use on children or infants	Skin peels, skin toner, face and body cream, cuticle softener, skin cleanser, skin improvers, shampoo	
Skin rashes; severe sloughing of the skin; ingestion of large doses can cause diarrhoea, nausea, vomiting and bleeding; see also Aluminium (E173) in section 1	Antiperspirant, deodorant, barrier cream	Waterproofing, fabric finishes, dye for furs

Names	Function	Code	
Aluminium chloride (salt of aluminium)	Deodorant agent	☹	
Aluminium chlorohydrate	Deodorant agent	☹	
Aluminium zirconium octachlorohydrate	Deodorant agent	☹	
Ambergris (derived from sperm whales; of ANIMAL origin)	Fixative Flavouring	☺☺	
Aminomethyl propanol (an alcohol made from nitrogen compounds)	Emulsifier	☺	

Potential Effects	Cosmetic Uses	Other Uses
Skin irritation; allergic reactions; reproductive and neurotoxicity; teratogenic; harmful to aquatic organisms; see also Aluminium (E173) in section 1	Lipstick, antiperspirant	
Contact allergic reactions; hair follicle infections; irritation of abraded skin; see also Aluminium (E173) in section 1	Antiperspirant, deodorant	
Harmful; contact allergic reactions; skin irritation; lung damage; granulomas; see also Aluminium (E173) in section 1	Non-aerosol antiperspirant, deodorant	
Ambergris is 80% cholesterol; no known adverse effects in humans	Perfumery	Flavouring for foods and beverages, cigarettes
Considered safe in cosmetic use up to 1% concentration; may cause skin irritation	Hairspray, shaving cream, cosmetic cream, deodorant	

Names	Function	Code	
2-Amino-4-nitrophenol and 4-Amino-2-nitrophenol (On a list of substances banned in the EU; on Canadian Hotlist)	Colourant	☹☹	
Aminophenol (m-,o-, p-)	Colourant	☹	
Ammonium carbonate	Neutraliser Buffer	☺?	
Ammonium chloride	Acidifier Buffer	☺?	
Ammonium cocoyl isethionate	Cleanser	☺	
Ammonium cocoyl sarcosinate (may be of ANIMAL origin)	Surfactant	☺?	
Ammonium hydroxide (solution of ammonia and water)	Buffering agent Denaturant	☺?	

Potential Effects	Cosmetic Uses	Other Uses
Animal carcinogens; possible human carcinogens; toxic to aquatic organisms	Orange-red and medium-brown hair dye	
Classified as toxic; possible risk of irreversible effects	Hair dye	
Skin rash on scalp, forehead and hands; contact dermatitis	Permanent wave lotion and cream	Fire extinguishers
Possible skin and eye irritation in some people; toxic to aquatic organisms	Bubble bath, hair bleach, shampoo	Batteries, dye, medicines
Considered safe in cosmetic use; may cause skin rashes	Soap, shampoo	
May be contaminated with carcinogenic nitrosamines; see Sarcosines	Shampoo, dentifrices	
Irritating to eyes and mucous membranes; may cause hair breakage; toxic by ingestion; harmful to the environment	Hair dye, hair straightener, barrier cream, mascara	Cigarettes, stain removers, detergents

Names	Function	Code	
Ammonium laureth sulphate	Surfactant	☺?	
Ammonium lauroyl sarcosinate (may be of ANIMAL origin)	Surfactant	☺?	
Ammonium lauryl sulphate	Surfactant Foaming agent	☺?	
Ammonium persulphate (ammonium salt)	Preservative Oxidiser	☺?	
Ammonium sulphate (ammonium salt)	Surfactant Cleanser	☹	
Ammonium thioglycolate (ammonium salt of thioglycolic acid; on Canadian Hotlist)	Antioxidant	☹	
Amyl acetate (banana oil; obtained from amyl alcohol)	Solvent Flavouring	☹	

COSMETIC INGREDIENTS

Potential Effects	Cosmetic Uses	Other Uses
May be contaminated with carcinogenic nitrosamines	Shampoo, bubble bath, hand wash	Dishwashing liquid, detergent
May be contaminated with carcinogenic nitrosamines; see Sarcosines	Shampoo, dentifrices	
Eye and skin irritation; repeated contact may dry the skin; may cause contamination with nitrosamines	Shampoo, bubble bath, liquid hand and body wash, toothpaste, bath gel	Dishwashing liquid, car wash detergent
Asthma; mucous membrane and skin irritation; brittle hair	Cosmetics, dye, skin lightener, soap	Detergents
Liver, neuro- and respiratory toxicity; dry and denatured hair	Permanent wave lotion	Tanning, filler in vaccines
Severe burns and blistering of the skin; hair breakage; cumulative irritant; severe allergic reactions; lethal to mice in large injected doses	Hair straightener, depilatories, permanent wave lotion	
Headache; fatigue; chest pain; CNS depression; neuro- and respiratory toxicity; mucous membrane irritation	Perfume, nail polish, nail polish remover	Banana flavouring in foods, perfuming shoe polish

Names	Function	Code	
Amyl dimethyl PABA (Padimate A)	UV absorber	☹	
Anethole (from anise oil and others)	Flavouring Denaturant	☺?	
Aqua (water)	Solvent	☺☺	
Ascorbic acid (vitamin C)	Antioxidant Preservative	☺☺	
Ascorbyl palmitate (derived form ascorbic acid)	Preservative Antioxidant	☺	
Azo dye (extract from coal tar or crude oil; see coal tar)	Colourant	☹☹	

Potential Effects	Cosmetic Uses	Other Uses
May cause sensitisation; increase breast cancer cell division; estrogenic; endocrine disruption; carcinogenic	Sunscreen preparations	
Hives; skin blistering and scaling; gum and throat irritation	Perfume, toothpaste, mouthwash	
No adverse health effects, provided water is purified so it is not contaminated with chlorine, sodium fluoride etc	Many cosmetic and personal care products	Canned and bottled foods and beverages
Vitamin C plays many beneficial roles in the body	Cosmetic cream, antiwrinkle products	
Some palmitates may cause contact dermatitis	Cosmetic cream and lotion	
Skin contact can cause hives, asthma; hay fever; allergic reactions; bladder cancer; may be absorbed through the skin	Non-permanent hair rinses and tints	Foods and beverages

Names	Function	Code
Balsam Peru (extract from South American tree)	Antiseptic	☺?
Barium sulphate	Depilatory agent	☺?
Barium sulphide (on Canadian Hotlist)	Opacifier Depilating agent	☹
Beeswax (from bees; of ANIMAL origin or may be synthetic)	Emulsifier Emollient	☺
Behentrimonium chloride	Preservative	☺?
Bentonite (white clay)	Thickener Emulsifier	☺?

Potential Effects	Cosmetic Uses	Other Uses
Skin irritation, stuffy nose; contact dermatitis; common sensitiser; may cross-react with benzoic acid and others	Cream hair rinse, face masks, perfume	Cigarettes
Often causes skin reactions, poisonous when ingested	Depilatories, cosmetics	
Skin rashes; chemical burns; never apply to inflamed skin; poisonous if ingested	Depilatories, hair relaxers, cosmetics	
Considered safe in cosmetic use; can cause mild allergic reactions and contact dermatitis	Lipstick, mascara, baby cream, eye make-up, foundation	Confectionery, soft drinks, chewing gum
See Quaternary Ammonium Compounds	See Quaternary Ammonium Compounds	
Inert and generally non-toxic; may clog skin pores inhibiting proper skin function; venous injection causes blood clots and possibly tumours	Facial masks; make-up	Colourant in wine

Names	Function	Code	
Benzaldehyde (synthetic almond oil)	Solvent Flavouring	☹☹	
Benzalkonium chloride (BAK; on Canadian Hotlist)	Preservative Detergent	☹	
Benzene (derived from toluene or gasoline; on Canadian Hotlist)	Solvent	☹☹	
Benzethonium chloride	Preservative Antistatic agent	☺?	

	Potential Effects	Cosmetic Uses	Other Uses
	Highly toxic; eye and skin irritation; allergic reactions; CNS effects; convulsions; kidney, liver, respiratory and neuro-toxicity; on NIH hazards list; harmful to aquatic organisms	Cosmetic cream and lotion, soap, perfume, dye	Flavouring in sweets, cordials and ice cream, cigarettes
	Toxic; eye and skin irritation; contact dermatitis; conjunct-ivitis; can be fatal if ingested	Shampoo, hair conditioner, mouthwash, eye lotion	Antiseptic and detergent in medicinal use
	Highly toxic; liver, endo-crine, immuno-, respiratory and neurotoxicity; skin rash and swelling; teratogen; carcinogen-ic; very toxic to aquatic organisms	Nail polish removers	Detergents, nylon, artificial leather, varnish, lacquer, oven cleaner, paint
	Endocrine toxicity; skin irritation; toxic to aquatic organisms; see Quaternary Ammonium Compounds	Cosmetics, feminine hygiene products	

Names	Function	Code	
Benzophenones (1–12) (a dozen or more different ones exist)	Flavouring Fixative UV absorber	☹	
Benzoyl peroxide (from benzoic acid; on Canadian Hotlist)	Bleaching and drying agent	☹	
Benzyl acetate (obtained from plants, especially jasmine)	Flavouring Solvent	☹	
Benzyl alcohol (constituent of jasmine, hyacinth and other plants; synthetically derived from petroleum or coal tar)	Solvent Preservative Denaturant	☹	
Benzyl carbinol (phenethyl alcohol)	Preservative	😐?	
Benzyl cinnamate (sweet Odour of Balsam)	UV absorber Additive	😐?	

Potential Effects	Cosmetic Uses	Other Uses
Hives; photoallergic reactions; contact sensitivity; toxic when injected; on NIH hazards list; harmful to aquatic organisms	Hairspray, soap, sunscreen, perfume	Flavourings for various foods
Skin irritation; toxic if inhaled; allergic reactions; corrosive; AVOID SKIN CONTACT	Cosmetics, artificial nail kits	
Vomiting; diarrhoea; eye and skin irritation; liver and neuro-toxicity; on NIH hazards list	Perfume, soap	Ice cream, baked goods, chewing gum
Headache; skin and mucous membrane irritation; neuro- and liver toxicity; contact dermatitis; on NIH hazards list; toxic to aquatic organisms	Perfume, hair dye, shampoo, nail varnish remover	Fruit flavourings for foods, fabric softener, cigarettes
Eye irritation; toxic if ingested; sensitiser; birth defects in rats; CNS injury in mice	Cosmetics, most rose perfume	Synthetic fruit flavouring in foods
Cinnamates can cause a stinging sensation in some people; on NIH hazards list	Cosmetics, perfume	Cigarettes, pesticides

Names	Function	Code	
Benzylhemiformal	Preservative	😐?	
Benzyl salicylate (synthetic)	UV absorber	😐?	
Betaglucans (found in oat fibre and barley)	Thickener	😊😊	
Beta-naphthol (from naphthalene from coal tar; on Canadian Hotlist)	Solvent	😦😦	
Beta hydroxy acids (BHA's; salicylic acid, beta hydroxy butanoic acid, tropic acid, trethocanic acid)	Exfoliant	😐?	
BHA	Preservative	😦😦	
BHT	Preservative	😦😦	
Biotin (water-soluble vitamin)	Texturiser Moisturiser	😊😊	

	Potential Effects	Cosmetic Uses	Other Uses
	Prolonged skin contact may be harmful	Cosmetics	
	Skin rash and swelling on exposure to sunlight	Sunscreen, perfume	Cigarettes
	No known adverse effects; may have beneficial effects	Facial powder, skin conditioner	
	Kidney damage, eye injury, convulsions, anaemia and death from ingestion; skin damage; contact dermatitis	Hair tonics, hair dye, skin peels, perfume	
	Photosensitivity; skin reactions especially if skin is dry or sensitive; changes skin pH; do not use on children	Exfoliant cream, skin peels, skin masks, moisturiser	
	See Butylated Hydroxyanisole (E320) in section 1	Cosmetics	Foods
	See Butylated Hydroxytoluene (E321) in section 1	Cosmetics, lipstick, eyeliner, baby oil	Foods, packaging materials
	No known adverse effects; beneficial effects on health	Cosmetic cream, hair conditioner	

Names	Function	Code	
Bismuth compounds (bismuth citrate, bismuth oxychloride etc)	Various	☹	
Bisphenol A	Hardener	☹	
Borax (sodium tetraborate)	Emulsifier Texturiser	☺?	
Boric acid (on Canadian Hotlist)	Anti-microbial	☹☹	
Bromates (calcium, potassium, sodium bromate)	Maturing agent	☹	

	Potential Effects	Cosmetic Uses	Other Uses
	Toxic effects include memory loss, convulsions, confusion, intellectual impairment; kidney and cardiovascular toxicity	Bleaching and freckle cream, nail polish, hair dye	
	Neurological diseases; learning difficulties; birth defects in mice; oestrogen mimic		Some plastic storage containers for foods and cosmetics
	See Sodium Tetraborate (E285) in section 1; harmful to aquatic organisms	Cold cream; shaving cream	Water softener, insecticide
	Gastrointestinal, liver, kidney, reproductive, blood and neurotoxicity; severe poisoning has occurred after ingestion and application to abraded skin	Baby powder, bath powder, eye cream, mouthwash, soap	Fungus control on citrus fruit
	Respiratory depression; skin eruptions; kidney dysfunction and failure; effects on the CNS	Permanent wave neutraliser	Used in making bread

Names	Function	Code	
Bromochlorophene (phenolic compound)	Preservative	☹	
2-Bromo-2-nitropropane 1,3-diol (Bronopol™, BNPD)	Preservative Solvent	☹	
5-Bromo-5-nitro-1,3-dioxane (Bronidox L)	Preservative	☹	
Bronidox L	Preservative	☹	
Bronopol	Preservative Solvent	☹	
Butane (derived from petroleum)	Propellant	☺?	
Butyl acetate (synthetic; derived from butane)	Solvent Flavouring	☹	
Butyl alcohol (synthetic; derived from butane)	Solvent Clarifier	☹	

COSMETIC INGREDIENTS

Potential Effects	Cosmetic Uses	Other Uses
Acutely toxic when ingested; see Hexachlorophene	Cosmetics	
Eye and skin irritation; liver toxicity; contact dermatitis; can produce carcinogenic nitrosamines and formaldehyde	Shampoo, mascara, eye make-up, liquid hand wash, nail polish, face cream	
Skin and eye irritation; can release formaldehyde; can form nitrosamines	Shampoo, mascara, eye make-up, liquid hand wash	
See 5-Bromo-5-Nitro-1,3-Dioxane	See 5-Bromo-5-Nitro-1,3-Dioxane	
See 2-Bromo-2-Nitropropane-1, 3-Diol	See 5-Bromo-5-Nitro-1,3-Dioxane	
See Butane (E943a) section 1	Aerosol cosmetics	Refrigerant
Toxic; can cause skin and eye irritation; conjunctivitis; irritation of respiratory tract	Perfume, nail polish remover, eye make-up, soap	Synthetic flavouring in foods, cigarettes
Contact dermatitis; dry skin; ingestion can cause mucous membrane irritation; drowsiness, headache, dizziness	Shampoo; nail polish, nail polish remover	Synthetic flavouring in foods, waxes, shellac, resin, cigarettes

Names	Function	Code	
Butylated hydroxyanisole	Preservative Antioxidant	😐😟	
Butylated hydroxytoluene	Preservative Antioxidant	😐😟	
Butylene glycol (synthetic)	Humectant Solvent	😐?	
Butyl myristate (from myristic acid and butyl alcohol)	Emollient	🙂	
Butylparaben (synthetic; ester of butyl alcohol and p-hydroxybenzoic acid)	Preservative	🙁	
Butyrolactone	Solvent	🙁	
C13-14 isoparaffin	Solvent	🙂	

	Potential Effects	Cosmetic Uses	Other Uses
	See Butylated Hydroxyanisole (E320) in section 1	Cosmetics	Foods
	See Butylated Hydroxytoluene (E321) in section 1	Lipstick, eyeliner, baby oil and lotion	Foods
	Not on the GRAS list of the FDA; ingestion may cause renal damage, vomiting, drowsiness, depression, kidney damage, coma and death; may be harmful to the environment	Hairspray, setting lotion	
	May cause skin irritation; some myristates can promote acne	Lipstick, face cream, nail polish, nail polish remover	
	Allergic reactions; skin irritation; see Parabens	Cosmetics, shampoo	
	Toxic; possible human carcinogen; on NIH hazards list	Nail polish, nail polish remover, cosmetics	Making poly-vinylpyrro-lidone
	Pure paraffin is thought to be harmless to the skin; impurities can cause eczema and irritation	Moisturiser, soap, shaving products, sunscreen	

Names	Function	Code	
Caffeine (obtained as a by-product of decaffeinated coffee)	Flavouring	☹	
Calcium acetate (synthetic; salt of acetic acid)	Emulsifier	☺?	
Calcium myristate (may be of ANIMAL origin)	Surfactant	☺	
Calcium silicate	Anticaking agent	☺?	
Calcium stearate (prepared from limewater)	Opacifier Colourant	☺	
Calcium sulphide (formed by heating gypsum with charcoal)	Depilating agent	☹	

Potential Effects	Cosmetic Uses	Other Uses
Liver, neuro-, gastrointestinal, kidney and musculoskeletal toxicity; teratogenic; NRC; on NIH hazards list	Flavouring in lipstick, helps other ingredients penetrate the skin	Liqueurs, cola soft drinks, chocolate, cigarettes
Allergic reactions; on NIH hazards list	Fragrances in cosmetics	Dyeing, tanning and curing skins
Considered safe in cosmetics; some myristates can promote acne	Cosmetics	
Practically non-toxic orally; irritation of lungs and respiratory tract; allergic skin reactions	Face powder	Baking powder, lime glass
Considered safe in cosmetic use; being reviewed for safety	Shampoo, hair conditioner	Paints, printing ink, pesticides
Toxic; skin and eye irritation; can cause allergic reactions; corrosive	Depilatories	Luminous paints

Names	Function	Code	
Calcium thioglycolate	Depilating agent	☹	
Calomel (mercurous chloride; banned from cosmetics in the EU and USA)	Bleaching agent	☹☹	
Camphor oil (banned in the USA for use as a liniment)	Preservative	☹	
Canthaxanthin (may be of ANIMAL origin)	Colourant (pink)	😐?	
Caprylic/capric/lauric triglyceride (may be of ANIMAL origin)	Emollient Solvent	☺	

	Potential Effects	Cosmetic Uses	Other Uses
	Harmful; skin problems on hands or scalp; haemorrhaging under skin; severe allergic reactions; thyroid problems in laboratory animals	Permanent wave lotion, cream depilatories	Tanning leather
	Teratogenic; mercury poisoning; persists in the environment; very toxic to aquatic organisms; see Mercury Compounds	Skin bleaches, freckle cream; 'beauty cream'	
	Spasms; convulsions; dizziness; liver and neurotoxicity; contact dermatitis; respiratory problems; foetal death	Hair tonic, after-shave and pre-shave lotion	Spice flavour in foods, embalming fluid, moth-balls
	Oral intake can cause loss of night vision; see also Cantha-xanthin (E161g) in section 1	Artificial tanning aids	Chicken feed to colour egg yoke
	Low toxicity, mild eye and skin irritation	Lipstick, bath oil, perfume, soap, hairspray	

Names	Function	Code	
Captan (derived from phenol)	Preservative	☹	
Carbitol	Humectant Solvent	☹	
Carbomer 934, 940, 941	Thickener Emulsifier	☺?	
Carboxymethyl cellulose (made from cotton by-products; may be GM)	Stabiliser Emulsifier	☺?	
Castor oil (from the seeds of the castor oil plant)	Plasticiser	☺?	
Catechol (phenol alcohol in catechu black from *Acacia catechu*, on a list of substances banned in the EU)	Modifier	☹☹	

Potential Effects	Cosmetic Uses	Other Uses
Immuno- and neurotoxicity; reproductive disorders; teratogenic; may be hazardous to aquatic and soil organisms	Soap, shampoo, cosmetics	Agricultural fungicide
Hazardous at concentrations over 5%; more toxic than polyethylene glycol (see)	Sunscreen, nail polish	
Allergic reactions; eye irritation; safety under review	Cosmetics, toothpaste	Industrial uses
Toxicity when used in cosmetics is unknown; caused cancer and tumours in some animal studies	Shampoo, hand cream, shaving cream, hair grooming aids	Ice cream, beverages, laxatives
Allergic reactions; ingestion can cause pelvic congestion and induce abortion	Lipstick, bath oil, shaving cream, nail polish, face masks	Embalming fluid, laxatives, lamp oil
Liver, cardiovascular, neuro- and immunotoxicity; contact dermatitis; teratogenic; carcinogenic; toxic to aquatic organisms	Hair colouring, skin care preparations	

Names	Function	Code	
Ceresin wax (brittle wax derived from the mineral ozokerite)	Thickener Antistatic agent	☺	
Cetalkonium chloride (derived from ammonia)	Preservative Antibacterial	☺?	
Ceteareths -3, -6, -12, -20, -25, -30, -33 (of ANIMAL origin)	Emulsifier Emollient	☺?	
Cetearyl alcohol (may be natural or synthetic; may be of ANIMAL origin)	Emulsifier Emollient	☺	
Cetearyl glucoside (synthetic oleochemical from coconut and corn; may be GM)	Emulsifier	☺	
Cetearyl palmitate (may be of ANIMAL origin)	Emollient	☺	
Ceteth -1, -2, -4, -6, -10, -20, -30	Emulsifier Surfactant	☺?	

	Potential Effects	Cosmetic Uses	Other Uses
	Considered safe in cosmetic use; may cause sensitisation in some people	Barrier cream, hair conditioner, cream rouge, lipstick	Waxed paper and cloth, dentistry
	Contact allergies; dry, brittle hair; ingestion can be fatal; see Quaternary Ammonium Compounds	Hair conditioner, deodorant cosmetics, antiperspirant	
	Skin dryness; allergic reactions; may be contaminated with the carcinogens 1,4-dioxane and ethylene oxide	Cosmetics, suntan products, shampoo, moisturiser, hair conditioner	
	May cause contact dermatitis and contact sensitisation in some people	Hair tints, lipstick, shampoo, suntan preparations	
	See Cetearyl Alcohol	Hand and body lotion	
	Some palmitates may cause contact dermatitis	Hand lotion	
	May be contaminated with the carcinogens 1,4 dioxane and ethylene oxide	Hair products, skin care preparations, moisturiser	Detergents

Names	Function	Code	
Cetrimonium bromide (synthetic)	Preservative	☹	
Cetrimonium chloride (synthetic)	Preservative	😐?	
Cetyl alcohol (synthetic oleochemical; may be of plant, ANIMAL, or petrochemical origin)	Emollient Emulsifier Opacifier	☺	
Cetyl lactate (may be synthetic; may be of ANIMAL origin)	Emollient	☺☺	
Cetyl myristate (may be synthetic; may be of ANIMAL origin)	Emollient	☺	
Cetyl octanoate (may be of ANIMAL origin)	Emollient	☺	
Cetyl palmitate (may be synthetic; may be of ANIMAL origin)	Emollient	☺	
Cetyl ricinoleate (may be of ANIMAL origin)	Emollient Solvent	☺	

	Potential Effects	Cosmetic Uses	Other Uses
	Ingestion can be fatal; can cause skin and eye irritation; reproductive effects; teratogenic; toxic to mice embryos	Shampoo, deodorant, skin cleaning products	
	See Quaternary Ammonium Compounds	Shampoo, hair conditioner	
	Considered to have a low toxicity orally and on the skin; may cause hives and contact dermatitis; skin disorders	Baby lotion, mascaras, foundations, deodorant, antiperspirants, shampoo	Laxatives
	No known toxicity or adverse reactions	Cosmetics	Pharmaceutical preparations
	No known toxicity; may promote acne in some people	Cosmetics	
	See Cetyl Alcohol	Cosmetic cream, lipstick	Pesticides
	Considered safe in cosmetic use; some palmitates can cause contact dermatitis	Eye make-up, skincare preparations	Manufacture of lubricants
	Considered safe in cosmetic use; may cause eye irritation	Tanning preparations	

Names	Function	Code	
Cetyl stearate (may be synthetic; may be of ANIMAL origin)	Emollient	☺☺	
Chloracetamide (synthetic)	Preservative	☺?	
Chloramine-t (synthetic)	Preservative Antiseptic	☺?	
Chlorhexidine (synthetic; on Canadian Hotlist)	Preservative Topical antiseptic	☹	
Chloroacetamide (on Canadian Hotlist)	Preservative	☹	
Chlorobutanol (chlorbutanol; acetone chloroform)	Preservative Antioxidant	☹	
p-Chloro-m-cresol	Preservative	☹	
2-Chloro-p-phenylene-diamine	Intermediate	☹☹	

Potential Effects	Cosmetic Uses	Other Uses
No known toxicity or adverse reactions	Skin conditioner in cosmetic products	
See Acetamide and Quaternary Ammonium Compounds	Cold cream, mud packs, shampoo, cleansing lotion	
Skin irritation; allergic reactions	Mouthwash, nail bleaches	
Contact dermatitis; respiratory and immunotoxicity; has caused anaphylactic shock	Liquid cosmetics, feminine hygiene spray, deodorant	
Allergic reactions; contact dermatitis; immunotoxicity	Cosmetics	
Acute oral toxicity; CNS depression; allergic reactions; harmful if inhaled; can be absorbed into the skin	Eye lotion, baby oil	Treating mastitis in cows
Caused kidney damage and adrenal tumours in male rats; unsafe in cosmetic products	Skin care and suntan cosmetic products	
See Phenylenediamine	Hair dye	

Names	Function	Code	
Chlorothymol (thymol derivative; phenolic compound)	Oral care agent Deodorant	☺?	
Chloroxylenol (PCMX; synthetic)	Preservative Antibacterial	☹	
Choleth -10-24 (polyethylene glycol ether of cholesterol)	Emulsifier	☺	
Chromium compounds	Colourant	☹	
CI (number) e.g. CI 12085 (colour index; inorganic colour listing in the EU; mostly synthetic coal tar/azo dye)	Colourant	☺?	
Cinnamyl alcohol (synthetic)	Flavouring Fragrance	☺	

Potential Effects	Cosmetic Uses	Other Uses
Combined with chlorine can cause mucous membrane irritation and skin rashes; may be absorbed via the skin	Mouthwash, hair tonic, baby oil	Topical anti-bacterial medication
Toxic by ingestion; liver and immunotoxicity; skin irritation; may be absorbed via the skin	Brushless shaving cream, shampoo, deodorant	Germicides, antifungal preparations
No known adverse reactions; safety is under review	Hand cream	
Dust inhalation can cause irritation and ulceration; lung cancer years after exposure; allergic reactions	Green eye make-up, green mascara	
Many can be harmful and cause skin, eye and mucous membrane irritation; see Azo Dye and Coal Tar	Hair dye	
May cause allergic reactions	Synthetic perfume, deodorant	Flavour in food, cigarettes

Names	Function	Code	
Cinoxate (cinnamic acid)	Flavouring UV absorber	☺?	
Citronella oil (extract from fresh grass)	Flavouring Fragrance	☹?	
Coal tar (contains creosol, quinoline, xylene, phenol, benzene, naphthalene and others)	Colourant	☹☹	
Cocamide DEA (semi-synthetic)	Emulsifier Surfactant	☹	
Cocamide MEA (synthetic)	Surfactant Emulsifier	☺?	
Cocamidopropyl betaine (synthetic)	Surfactant	☺?	
Cocamidopropyl dimethylamine	Antistatic agent	☺?	
Cocoamidopropyl hydroxysultaine	Surfactant Thickener	☺	

Potential Effects	Cosmetic Uses	Other Uses
Allergic skin rashes; photoallergic reactions	Sunscreen, perfume	
Asthma; skin rash; hay fever; stuffy nose	Soap, cosmetics, perfume	Insect repellent, food flavouring
Contact dermatitis; psoriasis; hives; photo-toxicity; acne; skin rash; breast, bladder and liver cancers; harmful to the environment	Shampoo, hair dye, facial cosmetics, hand and body lotion, toothpastes	Adhesives, insecticides, creosotes, phenols
Allergic skin rash; can contain DEA see Diethanolamine	Shampoo, bubble bath, shaving gel	Pet shampoo
Mild skin reactions in some people; vapour is highly toxic; may contain nitrosamines; harmful to the environment	Shampoo, hair conditioner	
Contact dermatitis; allergic reactions; eyelid rash	Soap, eye make-up remover, shampoo	
Contact allergies; contact dermatitis in some people	Hair conditioner	
May cause allergic skin rash; may contain nitrosamines	Shampoo, hair and skin cream	

Names	Function	Code	
Cocoa butter (theobroma oil; from roasted seeds of the cocoa plant)	Emollient Emulsifier	☺	
Coco-betaine (synthetic; from coconut oil)	Surfactant	☺	
Coco-polyglucose (synthesised; may be GM)	Surfactant	☺	
Coconut acids, oil and alcohols (from coconut kernals)	Surfactant Emollient Solvent	☺	
Collagen (of ANIMAL origin)	Biological additive	☺?	
Corn flour (may be GM)	Absorbent	☺	
Corn oil (may be GM)	Emollient	☺	
Cornstarch (may be GM)	Dusting powder	☺?	

Potential Effects	Cosmetic Uses	Other Uses
Softens and lubricates the skin; may cause allergic skin reactions and cosmetic acne	Soap, eyelash cream, rouge, nail whitener, lipstick	Sweet sauces, confectionery, suppositories
May cause skin rash in sensitive people	Shampoo, face and hand gel	
May cause skin irritation in sensitive people	Cosmetics	
May alleviate dry skin; may cause allergic skin rashes; eye and skin irritation	Shampoo, baby soap, massage cream	Margarine, chocolate, cigarettes
May form a film which can inhibit proper skin function; allergic reactions	Hand cream, moisturiser, cosmetics	
Used as a safer alternative to talc; when moist it can promote fungal and bacterial growth	Baby powder, face and bath powder	
May cause allergic skin reactions in some people	Cosmetic cream, toothpaste	
May cause allergic reactions; skin rashes; asthma; see Corn flour	Dusting powder	Demulcent medication

Names	Function	Code	
Coumarins (derived from tonka beans or made synthetically; banned in foods in the USA)	Fragrance Additive	☹	
p-Cresol (obtained from coal tar)	Preservative Flavouring	☹	
Crystalline silica	Abrasive	☹	
Crystallins (may be of ANIMAL origin)	Biological additive	☺☺	
Cyclohexylamine (synthetic)	Additive Buffer	☹	
Cyclomethicone (silicone derived from silica)	Solvent Antistatic agent	☺?	

Potential Effects	Cosmetic Uses	Other Uses
Allergic contact dermatitis; toxic by ingestion; photosensitivity; carcinogenic; teratogenic; on NIH hazards list	Acne preparations, soap, deodorant, hair dye, shampoo, sunscreen, perfume	Detergents
Skin burns; dermatitis; respiratory failure; blood, endocrine, kidney, liver and neurotoxicity; less toxic than phenol; toxic to aquatic organisms	Mouthwash, cosmetics	Synthetic nut and vanilla flavour in foods
Eye, skin and lung irritation when used in its dry form; carcinogenic	Blusher, lip pencils, facial powder	'Kitty' litter, cleanser, paints
Currently no known adverse effects in cosmetic use	Hair and skin care products	
Cardiovascular, respiratory, reproductive, immuno- and neurotoxicity; skin burns	Hairspray	
No known toxicity, but it coats the skin which may inhibit proper functioning	Hair conditioner, lipstick, deodorant, skin fresheners	Waterproofing, lubricants

Names	Function	Code	
D and C colours; **e.g. D & C red no. 6** (blue, brown, green, red, orange, violet and yellow)	Colourant	☺?	
DEA (diethanolamine)	Solvent Emulsifier	☹	
DEA cetyl phosphate (may be of ANIMAL origin)	Surfactant	☹	
DEA cocamide	Surfactant	☹	
DEA laureth sulphate (synthetic or semi- synthetic)	Surfactant	☹	
DEA lauryl sulphate (synthetic or semisynthetic)	Surfactant	☹	
Decyl alcohol (derived from liquid paraffin)	Antifoamer Fixative	☺	
Decyl myristate (may be of ANIMAL origin)	Emollient	☺	

Potential Effects	Cosmetic Uses	Other Uses
Most can cause health effects including skin rash, allergic reactions, asthma	Most cosmetics, including soap, lip-gloss, nail polish	
See Diethanolamine	See Diethanolamine	
May contain DEA; see Diethanolamine	See Diethanolamine	
See Cocamide DEA	See Cocamide DEA	
Harmful to aquatic organisms; see Diethanolamine and Quaternary Ammonium Compounds	Cosmetics, liquid soap, shampoo, hair conditioner	
Harmful to aquatic organisms; see Diethanolamine and Quaternary Ammonium Compounds	Cosmetics, liquid soap, shampoo, hair conditioner	
Low toxicity on the skin in animal testing	Cosmetics, perfume	Fruit flavouring in foods
Myristates can promote acne in some people	Skin conditioner in cosmetics	

Names	Function	Code	
Decyl oleate (may be of ANIMAL origin)	Emollient Emulsifier	☺	
Decyl polyglucose (decyl alcohol and glucose)	Surfactant	☺	
Dexpanthenol (may be of ANIMAL origin)	Anti-inflammatory	☺☺	
2,4-Diaminoanisole (synthetic)	Colourant	☹	
2,4-Diaminophenol (synthetic)	Colourant	☹	
Diazolidinyl urea (Germall II; of ANIMAL origin)	Preservative	☹	
Dibehenyldimonium chloride (synthetic)	Antistatic agent	☺?	
Dibenzothiophene (from thioxanthrone; banned in cosmetics in Italy)	Additive	☹	
Dibenzoylmethanes	UV absorber	☺?	

	Potential Effects	Cosmetic Uses	Other Uses
	May promote acne in some people; safety under review	Hand cream, suntan products	
	May cause skin irritation in sensitive people	Cosmetics	
	See Panthenol	See Panthenol	
	Allergic contact dermatitis; mutagenic; carcinogenic	Hair dye	
	See Phenylenediamine	Hair dye	
	Sensitiser; contact dermatitis; eye and skin irritation; may release formaldehyde; not readily biodegradable	Shampoo, hair conditioner, shaving gel, moisturiser	Pesticides, textile industry
	See Quaternary Ammonium Compounds	See Quaternary Ammonium Compounds	
	CNS disorders; blood pressure problems; blood disorders	Antidandruff shampoo, acne products	Psychopharmaceutical products
	Photoallergy; contact allergy	Sunscreen	

Names	Function	Code	
Dibromofluorescein (made by heating resorcinol with a naphthalene derivative)	Colourant	☹	
Dibromosalan (banned in cosmetics in the USA)	Antiseptic Fungicide	☹	
Dibucaine	Local anaesthetic	☺?	
Dibutyl oxalate (synthetic)	Chelating agent	☺?	
Dibutyl phthalate (from phthalic acid isolated from a fungus; banned in nail polish in the EU)	Film former Solvent	☹☹	
Dicetyldimonium chloride	Surfactant	☺?	
Dichloromethane (methylene chloride)	Solvent	☹	
Dichlorophene (crystals from toluene)	Anti-microbial	☹	

Potential Effects	Cosmetic Uses	Other Uses
Sensitivity to light; skin rash; skin and eye inflammation; respiratory and gastro-intestinal symptoms	Indelible lipstick	
Light sensitivity resulting in skin rash and swelling	Soap, cream, lotion, powder	Detergents
Highly toxic to rats when injected into the abdomen	Wax depilatories	
Oxalates are toxic	Products restricted to professional use	
Liver, kidney, reproductive and neuro-toxicity; abdominal pain; nausea; vertigo; contact allergic reactions; teratogen; carcinogenic; xenoestrogen; toxic to aquatic organisms	Perfume, nail polish, deodorant, antiperspirant	Insect repellent
See Quaternary Ammonium Compounds	Hair conditioner	
See Methylene Chloride	See Methylene Chloride	
Harmful; developmental and neurotoxicity; skin rashes; allergic reactions	Shampoo, antiperspirant, deodorant	

Names	Function	Code	
Dicocodimonium chloride	Surfactant	☺?	
Dicyclohexyl sodium sulfosuccinate	Surfactant	☺?	
Didecyldimonium chloride	Surfactant	☺?	
Diethanolamidooleam-ide DEA	Surfactant	☹	
Diethanolamine (DEA; see Nitrosamines)	Solvent Buffer	☹	
Diethylene glycol (made by heating ethylene oxide with glycol)	Humectant Solvent	☹☹	

Potential Effects	Cosmetic Uses	Other Uses
See Quaternary Ammonium Compounds	See Quaternary Ammonium Compounds	
See Quaternary Ammonium Compounds	See Quaternary Ammonium Compounds	
See Quaternary Ammonium Compounds	See Quaternary Ammonium Compounds	
See Quaternary Ammonium Compounds and DEA	See Quaternary Ammonium Compounds	
Skin and mucous membrane irritation; cardiovascular, kidney, gastrointestinal, liver and neurotoxicity; can combine with nitrosating agents to form the carcinogenic nitrosamine NDELA; hormone disruption; on NIH hazards list; harmful to aquatic organisms	Cosmetics, soap, shampoo, hair conditioner, bubble bath, moisturising cream, liquid soap	Detergents
Eye and skin irritation; ingestion can be fatal; blood, liver, kidney and neurotoxicity; teratogen; on NIH hazards list	Cosmetic cream, hairspray	Paracetamol elixirs

Names	Function	Code	
Diethyl phthalate (made from ethanol and benzene derivatives)	Solvent Fixative Denaturant	☹☹	
Dihexyl adipate (from adipic acid)	Emollient Solvent	☺	
Dihydroxyacetone	Colourant Humectant	☻?	
Diisopropanolamine	Acid-alkali adjuster	☻?	
Dimethicone (created from silica using the petrochemical methanol)	Antifoaming agent Emollient	☻?	

COSMETIC INGREDIENTS

	Potential Effects	Cosmetic Uses	Other Uses
	CNS depression; mucous membrane irritation; skin, liver, endocrine, respiratory and neurotoxicity; teratogen; may be hazardous to the environment, especially fish	Perfume, nail polish	Insect repellent
	Adipic acid has no known human toxicity but large oral doses are lethal to rats	Moisturiser, skin care products, make-up	
	Allergic contact dermatitis; lethal in rats when injected	Artificial tanning preparations	
	On NIH hazards list; can combine with nitrosating agents to form nitrosamines	Hair dye, permanent waves, tonics, hair grooming aids	Corrosion inhibitor
	Low toxicity; skin irritation; allergic reactions; caused tumours and mutations in laboratory animals	Cosmetics, skin conditioner	Household detergents, topical drugs

Names	Function	Code	
Dimethyl phthalate (phthalates are benzene derivatives)	Film former Solvent	☹	
Dioctyl phthalate (phthalates are benzene derivatives)	Film former Solvent	☹	
Dioctyl sodium sulfosuccinate	Wetting agent	☺	
1,4-Dioxane (created during the manufacturing process; can be removed from products by vacuum stripping; on Canadian Hotlist)	Contaminant	☹☹	
Dioxin (TCDD; highly toxic and carcinogenic contaminant)	Contaminant	☹☹	

	Potential Effects	Cosmetic Uses	Other Uses
	Phthalates linked with testicular cancer and cell mutations; neurotoxicity; can be absorbed through skin; teratogenic	Musk, calamine lotion, insect repellent	Pesticides
	Phthalates linked with testicular cancer and cell mutations; CNS depression; teratogenic; bioaccumulation may occur in seafood	Perfume, nail polish	Pesticides
	Considered to be safe as presently used in cosmetics	Hair styling products	
	Hormone disruption; oestrogen mimic; kidney, liver, neuro- and cardiovascular toxicity; lowered sperm counts; can penetrate human skin; stress related illnesses; teratogenic; carcinogenic	May be in products with ingredients having polyethylene glycol, eth, polyoxyethylene, oxynol, polyethylene or PEG in their names	Pesticides
	Cardiovascular, liver, neuro-, gastrointestinal, respiratory, immuno-, endocrine and kidney toxicity; mutagen; teratogenic; carcinogenic	May be present in processed foods, chlorine bleached paper, plastic lined cartons and cans	Released when plastic is burnt, newsprint, pesticides

Names	Function	Code	
Diphenyl methane (from methylene chloride and benzene)	Fragrance	☺?	
Diphenyl oxide (synthetic; from benzene)	Chelating agent	☺?	
Disodium lauryl sulfosuccinate	Surfactant	☺?	
Disodium oleamide sulfosuccinate (may be of ANIMAL origin)	Surfactant	☺?	
Distearyldimonium chloride	Antistatic agent	☺?	
DMAE (in anchovies and sardines; may be of ANIMAL origin)	Emollient Firming agent	☺☺	
DMDM hydantoin (derived from methanol)	Preservative	☹	

Potential Effects	Cosmetic Uses	Other Uses
Local skin irritation; skin reaction to sunlight (prickling, swelling, pigmentation)	Perfumed soap	
Vapour toxic if inhaled	Perfumery, perfumed soap	
May cause contamination with carcinogenic nitrosamines	Shampoo, body wash, bubble bath	Household detergent
May cause contamination with carcinogenic nitrosamines	Shampoo, body wash, bubble bath	Household detergent
See Quaternary Ammonium Compounds	See Quaternary Ammonium Compounds	
Supplemental DMAE is considered to have beneficial effects on health	Skin toner, face and eye cream, anti-aging cream	Oral supplements
Skin and eye irritation; allergic reactions; dermatitis; may release formaldehyde	Cosmetics, shampoo, mascara, cream conditioner	

Names	Function	Code	
Dodecylbenzene sulphonic acid	Surfactant	😐?	
Dodecylbenzenyltrimonium chloride	Surfactant	😐?	
Dodecylhexadecyltrimonium chloride	Antistatic agent	😐?	
Dodecylxylditrimonium chloride	Antistatic agent	😐?	
Dried egg yoke (of ANIMAL origin)	Colourant Protein	😐?	
Drometrizole (derived from benzene)	Solvent UV absorber	☹	
EDTA	Sequestrant Preservative	😐?	
Egg powder (of ANIMAL origin)	Protein	☺	

	Potential Effects	Cosmetic Uses	Other Uses
	Skin irritation and sensitisation; vomiting if ingested; toxic to aquatic organisms	Shampoo	Detergents
	See Quaternary Ammonium Compounds	See Quaternary Ammonium Compounds	
	See Quaternary Ammonium Compounds	Hair conditioner	
	See Quaternary Ammonium Compounds	See Quaternary Ammonium Compounds	
	Allergic reactions including hives, eczema, anaphylaxis	Cosmetics	Root beer, soups, coffee
	Determined not to be safe in cosmetic use by CIR Expert Panel; see Benzene	Nail polish, cosmetics	
	See Ethylene Diamine Tetraacetic Acid	See Ethylene Diamine Tetraacetic Acid	
	Harmless unless one is allergic to egg products	Shampoo, face masks, cream	

Names	Function	Code	
Elastin (may be of ANIMAL origin)	Biological additive	☺	
Emu oil (of ANIMAL origin)	Biological additive	☺	
EPO (Evening primrose oil)	Tonic	☺☺	
Ethanol (ethyl alcohol; from the fermentation of carbohydrates)	Solvent Antibacterial	☹	
Ethanolamines (mono, di and tri-forms)	Preservative Emulsifier	☹	
Ethoxyethanol (on Canadian Hotlist)	Solvent	☹	
Ethoxyethanol acetate (2-ethoxyethyl acetate)	Solvent	☹	

	Potential Effects	Cosmetic Uses	Other Uses
	Considered safe in cosmetic use; may coat the skin inhibiting proper function	Shampoo, hair conditioner, skin cream and lotion	
	Reported to have beneficial effects on health	Hand cream, cosmetics	
	See Evening Primrose Oil	See Evening Primrose Oil	
	Endocrine, cardiovascular, liver and neurotoxicity; dry skin; contact dermatitis; irritation	Toothpaste, mouthwash, nail polish, hair spray, perfume	Laundry detergent, cigarettes
	Irritating to lungs, skin and eyes; hair loss; sensitisation; may be contaminated with carcinogenic nitrosamines	Hair dye, cold permanent-wave lotion, soap	
	CNS depression; kidney damage; developmental, reproductive and neurotoxicity; can penetrate the skin	Cosmetics, nail polish, shampoo	
	Toxic, but less than ethoxyethanol; harmful to aquatic organisms; see Ethoxyethanol	Nail polish	

Names	Function	Code	
2-Ethoxyethyl-p-cinnamate	UV absorber	🙂?	
4-Ethoxy-m-phenylene-diamine sulphate	Colourant	☹️☹️	
Ethyl acetate	Solvent Flavouring	☹️	
Ethyl alcohol	Solvent	☹️	
Ethylenediamine (synthetic; on Canadian Hotlist)	Solvent pH control	☹️	
Ethylenediamine tetraacetic acid (EDTA)	Sequestrant Preservative Chelating agent	🙂?	

	Potential Effects	Cosmetic Uses	Other Uses
	See Cinoxate	See Cinoxate	
	See Phenylenediamine	See Phenylenediamine	
	Skin irritation; prolonged inhalation can cause kidney and liver damage; neurotoxicity, drying and cracking of the skin	Perfume, nail polish, nail polish remover	Synthetic flavour in foods, cigarettes, pesticides
	See Ethanol	See Ethanol	See Ethanol
	Toxic if inhaled or absorbed by the skin; severe skin and eye irritation; asthma; contact dermatitis; sensitisation; harmful to aquatic organisms	Thigh cream, cosmetics	Metal polish, pesticides
	Reported to have health benefits when used in chelation therapy; adverse effects can include asthma; skin and mucous membrane irritation; kidney damage; teratogen; on NIH hazards list; harmful to aquatic organisms	Hair dye, shower gel, shampoo, bar soap, face and hand gels	Oral supplements, carbonated beverages, dishwashing liquid, pesticides

Names	Function	Code	
Ethylene glycol	Solvent	☹	
Ethylene oxide (on Canadian Hotlist)	Humectant	☹☹	
Ethyl ester of PVM/PA copolymer	Film former	☺☺	
Ethyl methacrylate (ester of ethyl alcohol and methacrylic acid; on Canadian Hotlist)	Thickening agents	☹	
Ethyl myristate (ethyl alcohol and myristic acid; may be of ANIMAL origin)	Emollient Flavouring	☺	
Ethyl palmitate (may be of ANIMAL origin)	Emollient Flavouring	☺	

	Potential Effects	Cosmetic Uses	Other Uses
	CNS depression; immuno-, liver, neuro-, respiratory, gastrointestinal and kidney toxicity; contact dermatitis	Perfume, liquid soap, cosmetics	Insect repellent, antifreeze, car wax, shoe products
	Liver, gastrointestinal, neuro-, respiratory and kidney toxicity; headache; vomiting; spontaneous abortion; teratogenic; mutagenic; carcinogenic; harmful to aquatic organisms	Cosmetics, shampoo	Fumigant used on ground spices and other processed natural seasonings
	Considered safe as presently used in cosmetics	Hair setting preparations	
	Skin irritation; allergic reactions; neurotoxicity; allergic contact dermatitis; teratogen; AVOID SKIN CONTACT	Nail polish, artificial nails	
	Considered safe in cosmetic use; some myristates can promote acne	Cosmetics	Flavouring in foods, cigarettes
	Considered safe in cosmetic use; some palmitates can cause contact dermatitis	Cosmetics	Flavouring in foods, cigarettes

Names	Function	Code	
Ethylparaben (ethyl p-hydroxybenzoate; synthetic)	Preservative	☹	
Ethyl salicylate	Flavouring	☺?	
Ethyl thioglycolate (synthetic)	Depilatory agent	☹	
Eucalyptus oil (from the fresh leaves of the eucalyptus tree)	Local antiseptic	☺?	
Eugenol (obtained from clove oil)	Additive Fixative	☺?	
Euxyl K 400	Preservative	☺?	
Evening primrose oil	Tonic	☺☺	

Potential Effects	Cosmetic Uses	Other Uses
Allergic reactions; skin irritation; contact dermatitis; moderate potential for bio-accumulation; see Parabens	Cosmetics, make-up, shampoo, deodorant	
Allergic reactions, especially in people allergic to other salicylates	Perfume	Flavouring in foods, cigarettes
Thioglycolates can cause skin irritations, hair breakage, severe allergic reactions	Depilatories	
Can cause allergic reactions and skin irritation; large oral doses (1 tsp) can be fatal	Skin fresheners	Flavourings in foods, local antiseptic
Allergic reactions; vomiting and gastric irritation if ingested; liver and neurotoxicity	Perfume, dentifrices	Flavourings in foods
There have been reports of allergic reactions; allergic contact dermatitis?	Cosmetics, toiletries	
Believed to have beneficial health effects	Cosmetics	

Names	Function	Code	
Farnesol (found in star anise, cassia, citronella, rose, balsam and others)	Additive Flavouring	☺?	
Ferrous sulphate	Antiseptic Flavouring	☹	
Ficin (an enzyme found in the latex of tropical trees; may be GM)	Protein Digestant	☺?	
Fluorescein	Colouring	☹	
Fluoride (cumulative poison; classified as a contaminant by the USEPA; on Canadian Hotlist)	Oral care Preservative Insecticide	☹☹	

	Potential Effects	Cosmetic Uses	Other Uses
	Considered safe in current use; mildly toxic by ingestion; caused mutations in laboratory animals	Perfumery	Flavouring in foods, cigarettes
	Gastrointestinal, liver, kidney, cardiovascular and neurotoxicity; teratogenic; carcinogenic?	Hair dye, cosmetics	Flavouring in foods, treatment of anaemia
	Skin, eye and mucous membrane irritation	Used in cosmetics as a protein digestant	Cheese, to clot milk, meat tenderiser
	Lip inflammation; photosensitivity; respiratory and gastrointestinal symptoms	Indelible lipstick, nail polish	Dying wool, silk and paper
	Hypothyroidism; arthritis; osteoporosis; carpal tunnel syndrome; impaired brain function; birth defects; hip fractures; stress fractures; liver, kidney, musculoskeletal and neurotoxicity; dental and skeletal fluorosis; teratogenic; carcinogenic	Toothpaste, cosmetics, mouthwash, dentifrices	Many products containing water e.g. soft drinks, cordial, fruit juice, canned and bottled foods, public water supplies, dental treatments

Names	Function	Code	
Formaldehyde (gas derived from the oxidation of methyl alcohol; banned in cosmetics in some countries; on Canadian Hotlist)	Preservative Anti-microbial	☹☹	
Geraniol and geranyl compounds	Additive Flavouring	☺?	
Geranium oil (extract from plants)	Botanical additive	☺?	
Glutaral	Preservative Germicide	☹☹	
Glutaraldehyde (glutaral; synthetic; amino acid occurring in green sugar beets)	Preservative Germicide	☹☹	

	Potential Effects	Cosmetic Uses	Other Uses
	Eye, nose and throat irritation; coughing; nosebleeds; liver, respiratory, immuno-, skin, reproductive and neurotoxicity; nausea; contact dermatitis; rash; asthma; on NIH hazards list; teratogenic; carcinogenic	Mascara, nail hardener, nail polish, soap, hair restorer, shampoo, anti-aging cream, bubble bath, deodorant	Furniture polish, car wax, filler in vaccines, defoaming agents, animal feeds
	Allergic reactions; contact dermatitis; toxic if ingested	Perfume, shampoo, cosmetics	Chewing gum, cigarettes
	Contact dermatitis and skin irritation in some people; ingestion can be fatal	Tooth powder, dusting powder, perfume	Ointments
	See Glutaraldehyde	See Glutaraldehyde	See Glutaraldehyde
	Contact allergic reactions; contact dermatitis; immuno-, developmental, reproductive, skin and respiratory toxicity; nausea; headache; aches and pains; palpitations; mood swings; asthma; teratogenic; very toxic to aquatic organisms	Antiperspirant, hairspray, deodorant, setting lotion, waterless hand soap	Food flavouring; disinfectant used in hospitals and dentistry

Names	Function	Code	
Glycerin, glycerol (may be a by-product of soap manufacture; may be from plants, or of ANIMAL origin)	Humectant	☺	
Glyceryl distearate (from glycerin and stearic acid may be of ANIMAL origin)	Emulsifier Emollient	☺?	
Glyceryl myristate (may be of ANIMAL origin)	Emulsifier Stabiliser	☺?	
Glyceryl oleate (may be of ANIMAL origin)	Emulsifier Emollient	☺?	
Glyceryl PABA (may be of ANIMAL origin)	UV absorber	☺?	
Glyceryl stearate (may be of ANIMAL origin)	Emulsifier Emollient	☺?	
Glyceryl thioglycolate	Depilatory agent Reducing agent	☹	

	Potential Effects	Cosmetic Uses	Other Uses
	Considered non-toxic and non-allergenic; may cause skin to dry out in low humidity; skin irritation in some people	Hand cream, face masks, barrier cream	Various food uses
	May cause allergic reactions; contact dermatitis	Skin freshener, mascara, shampoo, cuticle softener	
	May cause contact dermatitis; may promote acne	Baby cream, face masks, hand lotion	
	May cause contact dermatitis and skin allergies	Cosmetic cream and lotion	
	May cause contact dermatitis and photosensitivity	Sunscreen	
	May cause skin allergies; contact dermatitis	Make-up, cuticle softener	
	Contact dermatitis; thioglycolates can cause hair breakage, skin irritations, severe allergic reactions	Permanent wave lotion, depilatories	

Names	Function	Code	
Glycolic acid (made synthetically from chloroacetic acid)	Buffer Exfoliant	☹?	
Guar hydroxypropyl-trimonium chloride	Antistatic agent	☹?	
Hectorite (clay containing lithium and magnesium silicates; constituent of bentonite)	Absorbent Antistatic agent	☺	
Hedera helix (extract from English Ivy)	Botanical Toning agent	☹?	
Heliotropin (piperonal; purple diazo dye)	Additive Flavouring	☹?	
Hemp seed oil (from the hemp plant)	Emollient	☺☺	
Henna (from the ground-up dried leaves and stems of a shrub)	Colourant (red)	☺	

Potential Effects	Cosmetic Uses	Other Uses
Mildly irritating to skin and mucous membranes; may cause sun sensitivity; exfoliative dermatitis; NRC	Skin peelers, exfoliants	Dying, brightening copper
See Quaternary Ammonium Compounds	See Quaternary Ammonium Compounds	
Considered safe in cosmetic use; dust can cause lung irritation	Hair bleaches, eyeliners, foundations	Pesticides
Can cause severe skin rashes; blistering; itching; contact dermatitis	Bath products, face and hand cream	
Allergic reactions; skin irritation; CNS depression on ingestion of large amounts	Perfume, soap	Cherry and vanilla food flavours
No known adverse effects in cosmetic use	Lip balms, skin moisturiser	
One of the safest hair dyes; may cause allergic skin rash, avoid use near the eyes	Hair dye, conditioner and rinse	

Names	Function	Code	
Hexachlorophene (prohibited in most cosmetic products in the EU and USA; on Canadian Hotlist)	Preservative	☹☹	
Hexylene glycol (synthetic)	Solvent Viscosity controlling agent	☹	
Hexylresorcinol (derived from petroleum)	Antioxidant Antiseptic	☹	
Homosalate (homomethyl salicylate)	UV absorber	☺?	
Hyaluronic Acid (natural protein found in the body; of animal origin)	Humectant Antistatic agent	☺☺	

Potential Effects	Cosmetic Uses	Other Uses
Multiple sclerosis; contact dermatitis; gastrointestinal, liver and neurotoxicity; birth defects; asthma; blindness; chloasma; allergic reactions; very toxic to aquatic organism; bio-accumulates in the food chain e.g. milk; possible long term environmental effects	Antiperspirant, deodorant, baby oil, shampoo, toothpaste, cold cream, baby powder	Washing fruit, detergents, animal products
Contact dermatitis; eye, skin and mucous membrane irritation; gastrointestinal, liver, neuro- and respiratory toxicity	Cosmetics	Pesticides
Severe gastrointestinal irritation; bowel, liver and heart damage; allergic reactions	Mouthwash, sunburn cream	Anti-worming medicine, antiseptic
Endocrine disruption; reports of poisonings when absorbed through the skin	Sunscreen	
Considered to have beneficial health effects	Skin moisturiser, eye cream, hair conditioner	Oral supplements

Names	Function	Code	
Hydrazine (from chloramine, ammonia and sodium hydroxide; on Canadian Hotlist)	Reducing agent	☹	
Hydrogen peroxide (made from barium peroxide and diluted phosphoric acid; on Canadian Hotlist)	Preservative Oxidising agent	☺?	
Hydrolysed protein (of ANIMAL origin; contains MSG)	Flavouring Flavour enhancer	☺?	
Hydrolysed vegetable protein (derived from whey, it contains 10–30% MSG; may be GM)	Flavour enhancer Antistatic agent	☺?	

Potential Effects	Cosmetic Uses	Other Uses
Toxic if inhaled, ingested or absorbed through the skin; kidney, liver, cardiovascular, immuno- and neurotoxicity; carcinogenic; teratogenic; very toxic to aquatic organisms	Cosmetics	
Generally recognised as safe as a preservative in cosmetics; corrosive to skin, eyes and respiratory tract when undiluted; may cause allergic reactions, headache; nausea; toxic to aquatic organisms	Mouthwash, skin bleach, toothpaste, cold cream, hair bleach	Cheddar and Swiss cheese, medicinal antiseptic and germicide
Can cause contamination with carcinogenic nitrosamines; see MSG (E621) in section 1	Cosmetics, shampoo and hair treatments	Animal feed
Concerns associated with HVP include decreased body weight, organ atrophy, behavioural over-activity and bladder and bowel incontinence	Hair care products	Canned tuna, soup, sauces, packet meals

Names	Function	Code	
Hydroquinone (a phenol that occurs naturally, but is usually made synthetically from benzene; on Canadian Hotlist)	Antioxidant Bleaching agent	☹☹	
p-Hydroxyanisole (derived from hardwood tar or made synthetically; on Canadian Hotlist)	Antioxidant	☹	
Hydroxyethylcellulose (made from cellulose using petrochemicals; may be GM)	Binder Film former	☺	
Hydroxymethyl glycinate	Preservative	☺?	
Hydroxymethyl-cellulose	Thickener Additive	☺	
Hydroxypropyl methylcellulose (made from cellulose using petrochemicals)	Film former	☺	

	Potential Effects	Cosmetic Uses	Other Uses
	Nausea, vomiting, delirium and collapse from ingestion; eye damage; contact allergy; contact dermatitis; sensitisation; liver toxicity; mutagen; very toxic to aquatic organisms	Freckle cream, suntan lotion, hair colouring	Pesticides
	Non-Hodgkin's lymphoma; skin de-pigmentation; ingestion can cause intestinal tract irritation and heart failure; eye and skin irritation	Permanent hair colour, lipstick	
	Considered safe in cosmetic use; adverse reactions rare	Shampoo, tanning, products, mascara, hand and body lotion	
	May release formaldehyde; see Formaldehyde	Cosmetics	
	Considered safe in cosmetic use; adverse reactions rare	Cosmetics, hair care products	
	Considered safe in cosmetic use; mild eye and skin irritation; allergic reactions	Bubble bath, hair care products, tanning preparation	

Names	Function	Code	
Imidazolidinyl urea (of ANIMAL origin)	Preservative	☹	
Isobutanol (isobutyl alcohol)	Solvent	☹	
Isobutyl alcohol	Solvent	☹	
Isobutyl myristate (may be of ANIMAL origin)	Emollient	☺	
Isobutyl palmitate (may be of ANIMAL origin)	Emollient	☺	
Isopropanol (isopropyl alcohol; derived from petroleum)	Solvent Antifoaming agent	☹	
Isopropanolamine (MIPA)	pH control Solvent	☹	
Isopropyl alcohol	Solvent	☹	

	Potential Effects	Cosmetic Uses	Other Uses
	Contact dermatitis; may release formaldehyde; see Formaldehyde	Baby shampoo, eye make-up, bath oil, moisturiser, rouge	
	Toxic by inhalation; skin and mucous membrane irritation; dermatitis; neurotoxicity	Shampoo, fragrances	Synthetic fruit flavourings, cigarettes
	See Isobutanol	See Isobutanol	See Isobutanol
	Myristates can promote acne in some people	Cosmetics	
	Palmitates can cause contact dermatitis in some people	Cosmetics	
	Dry and denatured hair, skin irritation; liver, respiratory, gastrointestinal, kidney and neurotoxicity; teratogenic	Hair colour rinse, hand lotion, after-shave lotion, nail polish	Antifreeze, room deodorisers, shellac, carpet cleaner, car wax
	Severe skin and eye irritation; contact allergy and dermatitis; may form nitrosamines	Cosmetic cream	Insecticides
	See Isopropanol	See Isopropanol	See Isopropanol

Names	Function	Code	
Isopropyl isostearate (may be of ANIMAL origin)	Emollient	☺	
Isopropyl lanolate (of ANIMAL origin)	Lubricant Emollient	☻?	
Isopropyl linoleate (may be GM)	Emollient	☹	
Isopropyl myristate (may be of ANIMAL origin)	Emollient Solvent	☹	
Isopropyl palmitate (may be of ANIMAL origin)	Emollient Preservative	☻?	
Isopropyl stearate (may be of ANIMAL origin)	Emollient Binder	☻?	
Isostearyl neopentanoate (may be of ANIMAL origin)	Emollient	☺	

	Potential Effects	Cosmetic Uses	Other Uses
	Considered safe in cosmetic use; skin irritation when undiluted; may promote acne; see Stearic Acid	Skin conditioner, skin cleanser	
	May cause skin sensitisation; safety is under review	Cosmetics, skin cream, lipstick	
	CIR Expert Panel concluded there is insufficient data to support safety in cosmetics	Skin conditioner in cosmetics	
	May significantly increase the absorption of the carcinogen NDELA; may promote acne; on NIH hazards list	Suntan lotion, bath oil, shampoo, hand lotion, deodorant	Pesticides
	Eye and skin irritation; allergic reactions; on NIH hazards list	Moisturiser, baby lotion, cologne, hair care products	Pesticides
	May cause skin irritation and allergic reactions	Skin conditioner	Pesticides
	Considered safe in cosmetic use; may promote acne	Eye make-up, foundations	

Names	Function	Code	
Isostearyl palmitate (may be of ANIMAL origin)	Surfactant Emollient	☺	
Isothiazolinone	Preservative	☹?	
Kaolin (China clay)	Anticaking agent Absorbent	☹?	
Kathon CG (methylisothiazolinone and methylchloroiso-thiazolinone)	Preservative	☹	
Keratin (of ANIMAL origin; on Canadian Hotlist)	Film former Additive	☺	
Lanolin; lanolin oil; lanolin wax (may be contaminated with pesticides; of ANIMAL origin)	Emulsifier Emollient	☹?	
Lard oil (of ANIMAL origin)	Emollient	☺☺	

Potential Effects	Cosmetic Uses	Other Uses
May be a sensitiser for people who suffer allergies; may cause contact dermatitis	Hand cream, shaving cream, soap, protective cream	
May cause allergic reactions and contact dermatitis	Cosmetics	
May inhibit proper skin function by clogging the pores; chronic inhalation can affect the lungs leading to fibrosis	Baby powder, bath powder, face powder, make-up	Making pottery, porcelain, bricks
Contact dermatitis; potent sensitiser; bacterial mutagen; skin cancer	Shampoo, cosmetics	Leather preservation
Considered safe for most people when used in cosmetics	Permanent wave solution, shampoo, hair rinse and conditioner	
Thought to be safe if uncontaminated; may cause allergic skin reactions, acne and contact dermatitis	Lipstick, mascara, nail polish remover, eye make-up, hair conditioner	Pesticides
Considered safe in cosmetic use	Shaving cream, soap	Chewing gum base

Names	Function	Code	
Latex (synthetic rubber)	Film former	☺?	
Lauralkonium chloride	Preservative	☺?	
Lauramide DEA (synthetic derivative of coconut oil)	Thickener Foam booster	☹	
Lauramide MEA (synthetic derivative of coconut oil)	Antistatic agent	😐?	
Lauramidopropyl betaine	Antistatic agent	☺?	
Lauramine oxide	Antistatic agent	☺?	
Lauroyl lysine (may be of ANIMAL origin)	Viscosity controlling agent	☺☺	

Potential Effects	Cosmetic Uses	Other Uses
Skin rash; allergic reactions; anaphylaxis; ingredients of latex compounds can be poisonous	Beauty masks	Chewing gum base, gloves, balloons, condoms
Can cause eye irritation; may form nitrosamines	Should not be in products containing nitrosating agents	
Itchy scalp; allergic skin reactions; dry hair; may contain DEA; see Diethanolamine	Shampoo, hair conditioner, bubble bath	Dishwashing detergent
May cause mild skin irritation; may contain DEA; see Diethanolamine	Shampoo, hair conditioner	Dishwashing detergent
See Quaternary Ammonium Compounds	See Quaternary Ammonium Compounds	
Can form carcinogenic nitrosamines	Hair care products	
Considered safe in cosmetic use	Facial powder	

Names	Function	Code	
Lauryl alcohol (derived from coconut oil)	Surfactant Emollient	☺?	
Lauryl sulphate (derived from lauryl alcohol)	Foam booster	☹?	
Lavender oil (from the fresh flowery tops of the lavender plant)	Fragrance	☺	
Lead acetate (made from lead monoxide and acetic acid)	Colourant	☹☹	
Linalool (extract from oils of lavender, bergamot and coriander)	Additive	☹?	
Linoleamide DEA (diethanolamine and linoleic acid)	Emulsifier	☹	

Potential Effects	Cosmetic Uses	Other Uses
Skin irritation; may promote acne	Perfume, shampoo	Detergents
Skin sensitisation; moderate toxicity by ingestion; may contain formaldehyde	Shampoo	
Considered to have beneficial effects on health; may cause allergic contact dermatitis; photosensitivity?	Shampoo, skin fresheners, mouthwash, perfume, dentifrices	Antiseptic oils, cream and lotion, cigarettes
Lead poisoning; liver, kidney and neurotoxicity; affects brain development in infants and children; carcinogenic; toxic to aquatic organisms, bio-accumulates in plants and animals; persists in the environment	Hair dye, hair colour restorer for men	Skin treatment in animals, printing colours
May cause allergic reactions; facial psoriasis; mildly toxic by ingestion; skin and eye irritation; may effect the liver	Perfume, cologne, perfumed soap, aftershave, hand lotion, hairspray	Flavouring in foods, cigarettes, fabric softener
Can contain DEA; see Diethanolamine	Should not be in products containing nitrosating agents	

Names	Function	Code	
Linoleamide MEA (mixture of ethanolamides of linoleic acid)	Antistatic agent	😐?	
Linoleic acid (from edible fats and oils; may be of ANIMAL origin)	Emulsifier Antistatic agent	☺	
Linseed oil (from flaxseed)	Emollient	😐?	
Magnesium laureth sulphate	Surfactant	😐?	
Magnesium myristate (magnesium salt of myristic acid)	Opacifier	☺	
Magnesium oleth sulphate (of ANIMAL origin)	Surfactant	😐?	
4-MBC (methyl-benzylidene camphor)	UV absorber	☹	

	Potential Effects	Cosmetic Uses	Other Uses
	May be irritating to the skin and eyes; may contain DEA; see Diethanolamine	Hair care products	
	No known adverse effects in cosmetics; nausea and vomiting if large amounts ingested	Cosmetics	Vitamins, digestive aids, cheese making
	Cosmetic acne; allergic reactions	Shaving cream, medicinal soap	Paint, varnish, linoleum
	May cause mild irritation to skin and eyes; may contain carcinogens 1,4 dioxane and ethylene oxide (see both)	Shampoo	
	Myristates may promote acne in some people	Cosmetics	
	May contain carcinogens 1,4 dioxane and ethylene oxide (see both)	Cosmetics	Detergents
	Endocrine disruptor; estrogenic; increased uterine activity in pre-pubescent rats	Sunscreen	

Names	Function	Code	
Menthol (may be natural or synthetic)	Flavouring	☺?	
Mercaptans (compounds with reduced sulphur bound to carbon)	Fragrance	☹☹	
Mercury compounds (prohibited in most cosmetic products in the USA; on Canadian Hotlist)	Preservative	☹☹	
Methacrylic acid (on Canadian Hotlist)	Primer	☹	
Methanol	Solvent	☹	
Methenamine (made from formaldehyde and ammonia)	Preservative Antiseptic	☹	

Potential Effects	Cosmetic Uses	Other Uses
Allergic reactions; skin irritation; concentrate toxic if ingested; on NIH hazards list	Skin fresheners, perfume, mouthwash	Chewing gum, cigarettes, pesticides
Highly toxic; skin irritation; allergic reactions; infections of hair follicles	Depilatories	
Extremely toxic; blood, liver, kidney, neuro-, respiratory and reproductive toxicity; autism; epilepsy; teratogenic; can be absorbed through the skin; mercury is very toxic to aquatic organisms; bio-accumulates especially in fish	Medicated soap, cosmetics, freckle cream, facemasks, hair tonic, eye preparations	Dye, paint, fungicides, plastics
Poisonous if ingested; skin and nail damage; inflammation; burns; infection; neurotoxicity	Artificial nail kits; nail products	
See Methyl Alcohol	See Methyl Alcohol	
Can release formaldehyde; nitrosamine precursor; skin irritation; skin rash	Deodorant cream and powder, mouthwash	Medicines

Names	Function	Code	
Methicone (silicone)	Antistatic Emollient	☺?	
Methoxyethanol (ethylene glycol ether; on Canadian Hotlist)	Solvent Fragrance	☹	
4-Methoxy-m-phenylenediamine (4-MMPD; on Canadian Hotlist)	Colourant	☹☹	
4-Methoxy-m-phenylenediamine sulphate (4-MMPD sulphate; on Canadian Hotlist)	Colourant	☹☹	
5-Methoxypsoralen (5-MOP; banned from cosmetics in the EU)	UV absorber	☹	
8-Methoxypsoralen (8-MOP; banned from cosmetics in the EU)	UV absorber	☹	

	Potential Effects	Cosmetic Uses	Other Uses
	See Dimethicone	Lipstick, blusher, mascara, aftershave	
	Developmental and reproductive toxicity; birth defects, on NIH hazards list	Nail polish, perfume, liquid soap, cosmetics	
	CIR Expert Panel concluded that it is unsafe as a cosmetic ingredient; see Phenylenediamine	See Phenylenediamine	
	CIR Expert Panel concluded that it is unsafe as a cosmetic ingredient; see Phenylenediamine	See Phenylenediamine	
	Increased risk of skin cancer; contact allergy; photoallergy; neurotoxicity; carcinogenic	Suntan accelerator, sunscreen	
	Contact allergy; photoallergy; liver and neurotoxicity; carcinogenic	Suntan accelerator, sunscreen	

Names	Function	Code	
Methoxysalen (8-methoxypsoralen)	UV absorber	☹	
4-Methoxytoluene -2, 5-diamine HCL (On a list of substances banned in the EU)	Fragrance Flavouring	☹☹	
Methyl acetate (occurs naturally in coffee)	Solvent	☺?	
Methyl alcohol (methanol; on Canadian Hotlist)	Solvent Denaturant	☹	
Methyl anthranilate (synthetic, from coal tar)	Fragrance Flavouring	☺?	
Methylchloroisothia-zolinone (on Canadian Hotlist)	Preservative	☺?	
Methyldibromo glutaronitrile	Preservative	☹	

	Potential Effects	Cosmetic Uses	Other Uses
	See 8-Methoxypsoralen	Suntan accelerator, sunscreen	
	See Toluene	Perfume	Flavouring in foods
	Neurotoxicity; skin dryness, chafing and cracking	Perfume, toilet waters	
	Eczema; dermatitis; cardiovascular, liver, respiratory, endocrine and neurotoxicity; teratogen	Shampoo	Antifreeze, ink, paint, varnish, shellac, paint stripper
	Skin irritation; on NIH hazards list; see Coal Tar	Perfume, suntan lotion	Food flavour, cigarettes
	May cause allergic reactions; contact dermatitis; mutagen? see Kathon CG	Shampoo, liquid hand and body wash, aftershave	Dishwashing liquid
	Considered unsafe for use in cosmetic products; allergic reactions; contact dermatitis; skin sensitisation	Hair conditioner, bubble bath, indoor tanning preparation	Dishwashing liquid

Names	Function	Code	
Methylene chloride (dichloromethane; on Canadian Hotlist)	Fragrance	☹☹	
Methyl ethyl ketone (MEK; synthetic; usually from butyl alcohol)	Solvent Fragrance	☹	
Methyl isobutyl ketone (MIBK)	Flavouring Fragrance	☹☹	
Methylisothiazolinone (on Canadian Hotlist)	Preservative	😐?	

	Potential Effects	Cosmetic Uses	Other Uses
	Nausea; dizziness; eye and skin irritation; dermatitis; neuro-, liver, cardiovascular, kidney, endocrine and respiratory toxicity; carcinogenic; teratogenic; environmental hazard; possible ground water contamination	Nail polish, hair conditioner, shampoo, hairspray, cleansing cream	Tablet coatings, anaesthetic in medicine, decaffeination of some coffees
	Irritating to eyes, skin and mucous membranes; CNS depression; headache; liver and neurotoxicity; dermatitis	Shampoo, hair conditioner, nail polish, perfume	Paint thinners, adhesives
	Hazardous by ingestion or inhalation; kidney, gastrointestinal, respiratory, liver and neurotoxicity; dermatitis; birth defects; carcinogenic	Perfume	Synthetic fruit flavouring in foods, solvent for cellulose and lacquer
	See Isothiazolinone and Methychloroiso-thiazolinone	Baby products, hand wash, shampoo	Dishwashing liquid

Names	Function	Code	
Methyl methacrylate (banned in the EU; on Canadian Hotlist)	Film former	☹☹	
Methyl methacrylate crosspolymer	Film former	☹	
Methylparaben (methyl p-hydroxyben-zoate)	Preservative	☹	
Methyl salicylate (Oil of Wintergreen; may be synthetic)	Flavouring Disinfectant	☹	
Mexenone (2-Hydroxy-4-methoxy-4'- methyl-benzophe-none)	UV absorber	😐?	

Potential Effects	Cosmetic Uses	Other Uses
Severe skin irritation; allergic reactions; contact dermatitis; liver, blood, respiratory, kidney, reproductive, neuro- and immunotoxicity; teratogenic; harmful to aquatic organisms	Nail polish, artificial nails	Medical and dental orthopaedic cement, adhesives
See Methacrylic Acid	Nail products	
May cause allergic reactions; contact dermatitis; see Parabens	Many cosmetic and personal care products	
Strong irritant to the skin and mucous membranes; blood, liver, neuro-, reproductive and respiratory toxicity; teratogen; harmful to aquatic organisms	Toothpaste, mouthwash, sunburn lotion	Flavour in foods, detergents, cigarettes
Photoallergy; hives; contact allergy; chronic actinic dermatitis; can mimic or exacerbate an illness; see Benzophenones	Sunscreen	

Names	Function	Code	
Mica (pulverised silicate minerals)	Opacifier Colourant	☺?	
Milk (may be contaminated with traces of pesticides, GMO's, antibiotics and hormones; of ANIMAL origin)	Emollient	☺?	
Mineral oil (white oil; petroleum derivative)	Emollient	☹	
Mixed fruit acids	Exfoliating agent	☹	
Monoethanolamine (MEA; ingredients ending in MEA)	Humectant Emulsifier	☺?	
Montan wax (derived from lignite)	Emulsifier	☺	

Potential Effects	Cosmetic Uses	Other Uses
May cause irritation and lung damage if powder inhaled; gastrointestinal and liver toxicity	Face powder, eye make-up, lipstick, shampoo, mascara	
May cause allergic reactions from mild to severe; in cosmetic use it can cause pimples and acne if not rinsed thoroughly from the skin	Bath preparations, facemasks, face wash	Hidden? (cream of rice, filled candy bars, macaroni, items with casein)
Can inhibit proper functioning of the skin; dry skin; teratogenic; kidney and neurotoxicity when untreated or mildly treated; may be phototoxic	Baby cream and lotion, lipstick, cold cream, eye cream	Used as a food additive in some countries
See Alpha Hydroxy Acids	See Alpha Hydroxy Acids	See Alpha Hydroxy Acids
Can cause skin and eye irritation; may cause carcinogenic nitrosamine formation	Soap, cosmetics	Detergents, paint stripper
Considered non-toxic in cosmetic uses	Lipstick, foundations	

Names	Function	Code	
Morpholine (prepared from diethanolamine; on Canadian Hotlist)	Emulsifier Surfactant	☹	
Moskene	Fragrance	☹	
Musk (dried secretion from a deer; of ANIMAL origin)	Fragrance	☺	
Musk ambrette (of ANIMAL origin; banned from cosmetics in the EU)	Fixative Flavouring	☹	
Musk moskene (of ANIMAL origin; banned from cosmetics in the EU)	Fragrance	☹	
Myristalkonium chloride (may be of ANIMAL origin)	Surfactant Preservative	☺?	
Myristamide DEA (may be of ANIMAL origin)	Viscosity control	☹	

	Potential Effects	Cosmetic Uses	Other Uses
	Skin, eye and mucous membrane irritation; kidney, liver, respiratory and neurotoxicity; see Diethanolamine	Various cosmetics	Coating on fresh fruit and vegetables
	See Musk Moskene	See Musk Moskene	
	Generally safe and non-toxic; can cause allergic reactions in some people	Perfume	Flavouring in foods
	Neurotoxic; photosensitivity; contact dermatitis; serious brain damage in animals	Cosmetic cream, aftershave lotion, soap, dentifrices	Food flavouring, detergents
	Can cause non-permanent hyper-pigmentation; pigmented contact dermatitis	Perfume, rouges	
	See Quaternary Ammonium Compounds	See Quaternary Ammonium Compounds	
	See Diethanolamine	See Diethanolamine	

Names	Function	Code	
Myristic acid (may be of ANIMAL origin)	Emulsifier	😐?	
Nanoparticles (see also Glossary)	UV absorber	😐?	
1-Naphthol (coal tar derivative)	Colourant	🙁	
Neem seed oil (from a tree native to India)	Antibacterial Antiviral	🙂	
Neomycin (antibiotic; antibiotics are banned from cosmetics in the EU)	Antibiotic	🙁🙁	
Niacinamide (specific form of vitamin B3)	Additive	🙂	

Potential Effects	Cosmetic Uses	Other Uses
Can cause skin irritation; mutations in laboratory animals	Shampoo, shaving cream and soap	Food flavour, cigarettes
May cause DNA damage; bio-accumulate; cause brain damage in aquatic species	Cosmetics, barrier cream, sunscreen	
Toxic by ingestion and skin absorption; very toxic to aquatic organisms; see Coal Tar	Hair dye, perfumery	Treatments for skin diseases
Improves dry skin, eczema, acne and dandruff; teratogen?	Skin cream, soap, lipstick, shampoo	Insect repellent
Can cause allergic reactions, photoallergy, kidney toxicity, may promote staph infections	May be used in some underarm deodorant	
Considered to have many beneficial effects on health	Hair conditioner, anti-aging products	Cereal flours

Names	Function	Code	
Nickel sulphate	Additive	☹	
Nitrites (sodium, potassium etc)	Preservative Colour fixative	☹☹	
Nitrobenzene (essence of mirabane; nitric acid and benzene; on Canadian Hotlist)	Fragrance Solvent	☹☹	
2-Nitro-p-phenylene diamine (derived from coal tar)	Colourant	☹☹	

	Potential Effects	Cosmetic Uses	Other Uses
	Skin rash; kidney, endocrine and immunotoxicity; vomiting if ingested; contact dermatitis	Hair dye, eye pencils, cosmetics, astringents	Mineral supplement, nickel plating
	May combine with amines found in the stomach, saliva, foods and cosmetics to form carcinogenic nitrosamines	Sodium nitrite is used as an anticorrosive in some cosmetics	Cured meats, matches, tobacco
	Cyanosis; drowsiness; headaches; nausea; reproductive, kidney, liver, respiratory and neurotoxicity; absorbed through the skin; teratogenic	Cheap scented soap	Making aniline a base for dye and drugs, shoe polish
	See Phenylenediamine	See Phenylenediamine	

Names	Function	Code	
Nitrosamines (toxic group of compounds formed when nitrites and nitrates combine with amines; e.g. NDELA may be found in cosmetics and shampoo; on Canadian Hotlist)	Contaminant	☹☹	
Nylon (synthetic)	Thickener Opacifier	☺	
Octyl dimethyl PABA (Padamate O)	UV absorber	☺?	
Octyl dodecanol (may be of ANIMAL origin)	Solvent Surfactant	☺☺	
Octyl methoxycinnamate	UV absorber	☺?	
Octyl palmitate (may be of ANIMAL origin)	Emollient	☺	

	Potential Effects	Cosmetic Uses	Other Uses
	Can cause many forms of cancer including liver, lung, mouth, stomach and oesophageal; liver damage; can pass through the skin; environmental effects not adequately investigated	Cosmetic products and shampoo with DEA, MEA or TEA compounds unless removed by the manufacturer	Found in air, tobacco smoke, pesticides, water, cured meats
	Generally regarded as safe in cosmetic use, may cause allergic reactions in some	Mascara, eye make-up, highlighter, lengthening mascara	
	May cause sensitisation; increase breast cancer cell division; estrogenic; endocrine disruption; carcinogenic	Sunscreen, make-up	
	Generally regarded as safe in cosmetic use	Hair conditioner, lipstick	
	Photoallergy and contact allergy; endocrine disruption	Sunscreen, lipstick, perfume, foundation	
	Generally regarded as safe; may cause cosmetic acne	Cold cream, shaving cream, lipstick	

Names	Function	Code	
Oleamide DEA (may be of ANIMAL origin)	Viscosity control	☺?	
Oleic acid (may be of ANIMAL origin)	Emollient Defoaming agent	☺	
Oleoyl sarcosine (may be of ANIMAL origin)	Antistatic agent Surfactant	☹?	
Oleth 2 – oleth 50 (of ANIMAL origin)	Emulsifier Surfactant	☹?	
Olive oil (obtained from ripe olives)	Emollient Emulsifier	☺	

	Potential Effects	Cosmetic Uses	Other Uses
	Hives; can cause carcinogenic nitrosamine formation; see Diethanolamine	Shampoo, bubble bath, lipstick, soap	
	Low oral toxicity; may cause mild skin and eye irritation; may promote acne	Soft soap, lipstick, cosmetics	Cigarettes
	Can cause mild skin irritation; sarcosines can enhance absorption of other ingredients through the skin and can cause nitrosamine contamination	Soap, cosmetics, lubricants, hair conditioner	Polishing compounds
	Some oleths cause allergic reactions; limited information	Range of cosmetics and personal care	
	Generally safe; may cause allergic reactions and acne	Shampoo, lipstick, soap, hair oil	Massage oil

Names	Function	Code	
Orange oil (from the fresh peel of the sweet orange)	Fragrance Flavouring	☺	
Orris absolute (from stems of the iris plant)	Fragrance	☺	
Orthophenylphenol (O-phenylphenol)	Anti-microbial	☹	
Oxybenzone (derived from isopropanol)	UV absorber	☺?	
Oxyquinoline sulphate (made from phenols)	Preservative	☹	
PABA (on Canadian Hotlist)	UV absorber	☺?	
Padimate A (amyl dimethyl PABA)	UV absorber	☹	

	Potential Effects	Cosmetic Uses	Other Uses
	Allergic reactions if hypersensitive; severe reactions to concentrated oil of orange	Perfumery, soap, cologne	Cigarettes, pesticides
	Generally safe; can cause allergic reactions	Perfume	
	Very toxic; mutagenic; skin irritation; carcinogenic	Cosmetics	Disinfectant spray
	Photosensitivity; chronic actinic dermatitis; contact allergy	Sunscreen	
	See Phenol	Cosmetics	
	See Para-aminobenzoic Acid	See Para-aminobenzoic Acid	See Para-aminobenzoic Acid
	See Amyl Dimethyl PABA	Sunscreen	

Names	Function	Code	
Padimate O (octyl dimethyl PABA)	UV absorber	☹	
Panthenol (may be of ANIMAL origin)	Antistatic agent	☺☺	
Papain (from papaya; may be GM)	Additive	☺	
Para-aminobenzoic acid (found in vitamin B complex; on Canadian Hotlist)	UV absorber	😐?	
Parabens (butyl, ethyl, methyl, propyl etc; synthetic; esters of hydroxybenzoic acid)	Preservative	☹	
Paraffin (from petroleum, wood, coal or shale oil)	Emollient Viscosity control	😐?	
PBSA (2-phenylbenzimidazole-sulphonic acid)	UV absorber	😐?	

	Potential Effects	Cosmetic Uses	Other Uses
	See Octyl Dimethyl PABA	Make-up, sunscreen	
	Considered to have beneficial health effects	Hair care products, cosmetics	Digestive aid
	Considered to have health benefits; skin irritation	Skin cream, skin scrubs, skin masks	Meat tenderiser
	Helps prevent UV damage to skin and hair; photosensitivity; contact dermatitis; eczema; increase risk of skin cancer?	Sunscreen, sunburn lotion, shampoo, hair conditioner	Treatment for arthritis
	May cause allergic reactions; endocrine disruption; contact dermatitis; possible increase in the risk of breast cancer; toxic in animals by ingestion	Many cosmetic and personal care products	Various processed foods
	Pure paraffin is non-toxic to the skin; impurities can cause skin irritation and eczema	Lipstick, mascara, eyelash cream	Pesticides
	May increase the risk of skin cancer; skin damage	Sunscreen	

Names	Function	Code	
Pectin (found in roots, stems and fruits of plants)	Thickener Binder	☺☺	
Peg compounds e.g. PEG-20 myristate (polyethylene glycols or polymers of ethylene oxide)	Solvent Emulsifier	☹	
Peppermint oil (from the plant *Mentha piperita*)	Flavouring	☺?	
Petrolatum (white) (petroleum jelly; from petroleum)	Emollient Antistatic agent	☹	
Phenol (carbolic acid; derived from coal tar)	Preservative Denaturant	☹☹	
Phenoxyethanol (derived from phenol and ethylene oxide)	Preservative	☺?	

	Potential Effects	Cosmetic Uses	Other Uses
	Considered to have beneficial effects on health	Toothpastes, hair setting lotion, barrier cream	Foods, anti-diarrhoeal medicine
	Can be contaminated with 1,4-dioxane, ethylene oxide, lead and arsenic; see Polyethylene Glycols	Many cosmetic and personal care products	Manufacture of surfactants
	Can cause allergic reactions; hay fever; skin rash; allergic contact dermatitis	Toothpaste, shaving lotion	Foods and beverages, cigarettes
	Allergic skin reactions; skin discolouration; may contain the cancer causing contaminants poylcyclic aromatics	Wax depilatories, cold cream, eye make-up, lipstick	Glazing agent on some foods, pharmaceuticals
	Respiratory, cardiovascular, kidney, liver and neurotoxicity; paralysis; rash; nervous disorders; carcinogenic; toxic to aquatic organisms	Mouthwash, hand lotion, sunburn lotion, soap, shaving cream	Disinfectants
	Mild allergic skin rashes in sensitive people; concentrated solution can cause headache, nausea, renal failure	Shampoo, liquid soap, bubble bath, cosmetics, perfume	Insect repellent, antifreeze, filler in vaccines

WHAT'S REALLY IN YOUR BASKET?

Names	Function	Code	
Phenylalanine (essential amino acid; found in eggs, legumes, dairy products etc)	Artificial sweetener Antistatic agent	☺?	
Phenylbenzimidazole sulphonic acid	UV absorber	☹	
Phenylenediamine (m-,o-,p-) (m-phenylenediamine & o- phenylenediamine are on a list of substances banned in the EU)	Colourant	☹☹	
Phenylmercuric acetate	Preservative Fungicide	☹☹	
Phthalates (chemical compounds used in making PVC plastics; cosmetics; pesticides etc)	Film former Solvent	☹☹	

Potential Effects	Cosmetic Uses	Other Uses
Sufferers of phenylketonuria (PKU), melanoma or cirrhosis need to restrict intake; PKU, if not detected early can lead to mental deterioration in children	Hair conditioner	Some artificial sweeteners, cigarettes
See PBSA	See PBSA	
Eczema; asthma; skin rash; gastritis; contact dermatitis; blindness; cancer; death; very toxic to aquatic organisms	Most home and beauty salon hair dye, eyelash dye	
Allergic reactions; skin irritation; very toxic internally; very toxic to aquatic organisms; bio-accumulates in the food chain e.g. water organisms, fish, crustacea, birds	Mascara, shampoo	Paint
Kidney, reproductive, liver, endocrine and neurotoxicity; mutagen; carcinogenic; teratogenic; endocrine disruption; hazardous in the environment	Nail polish, hairspray, soap, shampoo	Production of PVC plastics, pesticides

WHAT'S REALLY IN YOUR BASKET?

Names	Function	Code	
Piperonal (purple diazo dye made from oxidation of piperic acid)	Additive Flavouring	☹	
Polyacrylamide (polymer of acrylamide monomers)	Thickener Film former	☹	
Polyethylene (may be contaminated with the carcinogen 1,4-dioxane)	Binder Antistatic Stabiliser	☺?	
Polyethylene glycols (may be contaminated with the carcinogen 1,4-dioxane)	Binder Solvent	☹	
Polyoxyethylene compounds (may be contaminated with the carcinogen 1,4-dioxane)	Emulsifier	☹	
Polypropylene glycol	Humectant	☹	
Potassium bromate (on Canadian Hotlist)	Antiseptic Astringent	☹	

	Potential Effects	Cosmetic Uses	Other Uses
	Skin rash; skin irritation; CNS depression; marking of the lips; on NIH hazards list	Soap, lipstick, perfume	Flavourings in foods, cigarettes
	CNS paralysis; highly toxic and irritating to skin; can be absorbed through the skin	Moisturising cream, nail polish, tanning cream, make-up	Adhesives, plastics, pesticides
	No known skin toxicity; large doses caused cancer in rats; ingestion of large doses can cause liver and kidney damage	Hand lotion, skin fresheners, suntan products, underarm deodorant	Chewing gum, sheets for packaging
	Skin and eye irritation; kidney and immunotoxicity; may contain traces of lead and arsenic	Barrier cream, lipstick, antiperspirant, baby products	Pharmaceutical ointments, oven cleaners
	Can cause sensitivity reactions; eye and skin irritation	Hand cream, hand lotion	Air freshener
	See Propylene Glycol	Liquid make-up	Pesticides
	Inflamed and bleeding gums; skin and eye irritation; nausea; very toxic when ingested	Toothpaste, mouthwash	Improving additive in bread

Names	Function	Code	
Potassium carbonate (inorganic salt of potassium)	Buffer	😐?	
Potassium chlorate (synthetic)	Oxidising agent	🙁	
Potassium hydroxide (caustic potash)	Emulsifier Buffer	😐?	
Potato starch (flour prepared from potatoes)	Emollient	🙂	
PPG compounds e.g. PPG-5-laureth-5	Various	🙁	
Propylene glycol (synthetic; from petroleum)	Humectant Solvent	🙁	

	Potential Effects	Cosmetic Uses	Other Uses
	Can cause dermatitis of the scalp, forehead and hands	Freckle lotion, shampoo, soap	
	Gum inflammation; dermatitis; intestinal and kidney irritation; may be absorbed through the skin	Toothpaste, freckle lotion, mouthwash, gargle	Bleach, fireworks, pesticides, matches
	Skin irritation and nail damage in cuticle removers; may cause skin rash and burning; hazardous to aquatic organisms	Liquid soap, barrier cream, hand lotion, cuticle removers	Household cleaners, button batteries
	Generally regarded as safe; may cause allergic skin reactions and stuffy nose	Dry shampoo, baby powder, dusting powder	
	See Propylene Glycol and Ethylene Oxide	Cosmetics	
	Contact dermatitis; lactic acidosis; skin rashes; dry skin; respiratory, immuno- and neurotoxicity; delayed contact allergy; increased absorption of other substances	Foundation cream, mascara, lipstick, baby lotion, suntan lotion, cold cream	Foods, cigarettes, pesticides, antifreeze

Names	Function	Code	
Propylene glycol alginate	Stabiliser Binder	☺?	
Propylparaben (propyl-p-hydroxybenzoate; synthetic)	Preservative	☹	
Psoralen (derived from a plant)	UV absorber	☺?	
Pumice (lightweight porous volcanic rock)	Abrasive cleanser	☺	
Pycnogenol (blend of bioflavonoids)	Antioxidant	☺☺	
Pyrocatechol (coal tar derivative; on Canadian Hotlist)	Antiseptic Oxidizer	☹	
Pyrogallol (a phenol; on Canadian Hotlist)	Antiseptic Colourant	☹	

Potential Effects	Cosmetic Uses	Other Uses
See Propane-1,2- diol Alginate (E405) in section 1	Cosmetics	Foods
Skin irritation and allergic reactions; contact dermatitis; photosensitivity; on NIH hazards list; see Parabens	Shampoo, beauty masks, nail cream, foundation cream, baby cream	Foods
Photodermatitis; photosensitivity	Sunscreen, perfume	Treatment of vitiligo
Generally regarded as safe; can cause irritation on dry, sensitive skin	Toothpaste, hand cleansing pastes, skin cleanser	
Considered to have beneficial effects on health	Anti-aging products	Chewing gum, supplements
Contact dermatitis; eczema; kidney and liver toxicity; carcinogenic	Hair dye, blonde-type dye, skin care preparations	Photography, dying furs
Skin sensitisation; skin rash; ingestion can cause kidney and liver damage; circulatory collapse; mutagen; teratogen; harmful to aquatic organisms	Permanent hair dye, skin care preparations	Anti-microbial, used medically to soothe irritated skin

Names	Function	Code	
Quaternary ammonium compounds (QUATS; synthetic derivatives of ammonium chloride)	Various	☺?	
Quaternium-15 (may break down to, or release formaldehyde)	Preservative	☹	
Quaternium-26 (may be contaminated with pesticides and DEA)	Surfactant Antistatic agent	☹	
Quaternium-18 hectorite	Viscosity controller	☺?	
Quercetin (type of bioflavonoid)	Colourant Antioxidant	☺	
Quillaja extract (extract from the bark of a tree in South America)	Surfactant	☺	

Potential Effects	Cosmetic Uses	Other Uses
All QUATS can be toxic depending on dose and concentration; contact dermatitis; eye and mucous membrane irritation; serious hypersensitivity and anaphylactic shock rarely	Aerosol deodorant, antiperspirant, hand cream, mouthwash, shampoo, lipstick, after-shave lotion	Medical sterilisation of mucous membranes, all-purpose cleaner
Contact dermatitis; allergic reactions; eye irritation; skin rash; sensitisation	Cosmetics, shampoo, hair conditioner	
Eye irritation; contact dermatitis; carcinogenic; see Diethanolamine (DEA)	Products giving sheen to hair	
See Quaternary Ammonium Compounds and Hectorite	Skin care products, suntan gels	
Considered to have beneficial health effects; may cause allergic reactions; on NIH hazards list; teratogenic	Dark brown shades of hair dye	Food additives, dyeing artificial hairpieces, supplements
Generally regarded as safe; large oral doses are toxic; may cause local irritation	Shampoo, skin cleanser, soap	Flavours for foods and beverages

Names	Function	Code	
Quinine (alkaloid from the bark of a South American tree)	Anaesthetic Flavouring	😐**?**	
Quinoline (coal tar derivative; may be of ANIMAL origin)	Solvent Colourant	🙁	
Resorcinol (derived from resins or may be synthetic)	Preservative Antiseptic Colourant	🙁	
Retinol (vitamin A; may be of ANIMAL origin; on Canadian Hotlist)	Preservative Additive	🙂	
Retinyl palmitate (ester of vitamin A; may be of ANIMAL origin; on Canadian Hotlist)	Texturiser Additive	🙂	

	Potential Effects	Cosmetic Uses	Other Uses
	Large or long-term dosages can cause headaches, skin rashes, intestinal cramps, tinnitus; cardiovascular and liver toxicity; teratogenic	Hair tonics, sunscreen preparations	Tonic water, 'bitter lemon' drinks, cold and headache remedies
	Psoriasis; dermatitis; gastrointestinal, liver, respiratory and neurotoxicity; carcinogenic; may be hazardous to the environment, especially fish	Manufacture of cosmetic dye	Preservative for anatomical specimens
	Eye and eyelid inflammation; dizziness; restlessness; endocrine disruptor; immuno-, liver, cardiovascular, neurotoxicity; harmful to aquatic organisms	Antidandruff shampoo, hair dye, lipstick, hair tonic	Tanning, explosives, printing textiles
	Considered to have beneficial health effects; excess levels can cause yellow skin, birth defects and liver toxicity	Massage cream and oils, skin care preparations	Topical acne treatments
	Considered to have beneficial health effects; safe in cosmetic use up to 1% concentration; contact dermatitis	Cosmetic cream, shaving cream; make-up; suntan products	

Names	Function	Code	
Rice starch (from pulverised rice grains; may be GM)	Emollient	☺?	
Ricinoleamide DEA (synthetic or semisynthetic)	Antistatic agent	☹	
Ricinoleic acid (from castor beans)	Emollient Emulsifier	☺?	
Rose hips oil (from rose hips)	Botanical	☺☺	
Rosemary extract (from an evergreen shrub)	Fragrance Flavouring	☺	
Rosin (obtained from pine trees)	Viscosity control	☺?	
Royal bee jelly (of ANIMAL origin)	Biological additive	☺	
Saffron (dried stigma of the crocus plant)	Colourant Flavouring	☺	

	Potential Effects	Cosmetic Uses	Other Uses
	Allergic reactions; can clog skin pores inhibiting proper skin function; acne	Baby powder, face powder	Foods
	Can contain DEA; see Diethanolamine	Cosmetics	
	Allergic reactions; dermatitis; on NIH hazards list	Soap, lipstick	Contraceptive jelly
	Considered to have beneficial effects on the skin	Skin cream, sun care products	Natural food flavouring
	Considered to have beneficial health effects; may cause photosensitivity	Bubble bath, skin cream, shampoo	Natural food flavouring
	May cause contact allergies; eyelid dermatitis; asthma	Soap, mascara, wax depilatories	Chewing gum, varnishes
	Considered by some researchers to have beneficial health effects	Cosmetics	
	Generally regarded as safe; may have beneficial health effects; anaphylaxis	Perfumery, cosmetics	Flavour in food and beverages, marking ink

Names	Function	Code	
Safrole (toxic component of volatile oils such as nutmeg and star anise; on Canadian Hotlist)	Fragrance Flavouring	☹☹	
Salicylates (salts of salicylic acid-benzyl, amyl, methyl, phenyl; found in fruits and vegetables)	Flavouring	☺?	
Salicylic acid (may be derived by heating phenol with carbon dioxide; one of the beta hydroxy acids; on Canadian Hotlist)	Preservative Antiseptic	☹	
Sarcosines and sarcosinates (found in starfish and sea urchins or formed from caffeine; may be of ANIMAL origin)	Surfactant	☺?	
Sassafras oil (volatile oil from sassafras plant; 80% safrole)	Fragrance Flavouring	☺?	

Potential Effects	Cosmetic Uses	Other Uses
Liver, kidney, reproductive and neurotoxicity; on NIH hazards list; carcinogenic; teratogenic	Cheap soap and perfume	Beverage flavouring?
Allergic reactions in people sensitive to aspirin; hyperactivity; kidney, cardiovascular and neurotoxicity; asthma	See Methyl Salicylate	Ice cream, jam, cake mixes, chewing gum, antiseptics
Large amounts absorbed can cause vomiting, abdominal pain, acidosis and skin rash; allergic reactions; dermatitis; teratogenic; aspirin-sensitive people should avoid	Skin softener, facemasks, make-up, hair dye remover, deodorant, suntan lotion	Food products, fungicide, topical treatment for acne
Non-irritating and non-sensitising; can cause formation of nitrosamines; can enhance penetration of other ingredients through the skin; see Nitrosamines	Shampoo, soap, dentifrices, lubricating oils	Dishwashing liquids
Dermatitis in sensitive people; unsafe in foods unless safrole-free; see Safrole	Perfume, soap, dentifrices	Flavouring in foods, topical antiseptic

Names	Function	Code	
Selenium sulphide (on Canadian Hotlist)	Antidandruff agent	☹	
Shea butter (from fruit of the karite tree)	Emollient Emulsifier	☺☺	
Silver nitrate	Colourant	☹	
Sodium alpha-olefin sulfonates	Cleanser	😐?	
Sodium carbonate (soda ash)	Buffer Oxidising agent	😐?	
Sodium chloride (common table salt)	Preservative Viscosity control	😐?	
Sodium cocoyl sarcosinate (may be of ANIMAL origin)	Surfactant	😐?	

	Potential Effects	Cosmetic Uses	Other Uses
	Skin irritation; dryness of hair and scalp; liver toxicity; severe eye irritation; carcinogenic	Medicated antidandruff shampoo	Treatment for tinea versicolour
	Softens and moisturises skin; no known toxicity	Moisturiser, lipstick, lip balm, suntan gel	
	Poisonous; caustic and irritating; skin sensitivity; allergies; very toxic to aquatic organisms	Metallic hair dye	
	May cause eye and skin irritation and sensitisation; foetal abnormalities in animals	Shampoo, bath and shower products	
	Breathing difficulty, abdominal pain, collapse from ingestion; liver toxicity; can cause scalp, forehead and hand rash	Shampoo, vaginal douches, soap, permanent wave lotion, bath salts	Dishwashing liquid, cigarettes, pesticides
	Can be irritating and corrosive to skin and mucous membranes; dry skin; skin rash; teratogenic	Shampoo, liquid hand wash, bubble bath, mouthwash	Butter, meats, cigarettes
	See Sarcosines	Shampoo, hand and body wash	

Names	Function	Code	
Sodium cocoyl isethionate	Surfactant	☺	
Sodium fluoride (on Canadian Hotlist)	Preservative Oral care	☹☹	
Sodium hydroxide (caustic soda)	Emulsifier Alkali	☹?	
Sodium hydroxymethyl glycinate	Preservative	☹?	
Sodium lauraminopropionate	Surfactant Antistatic	☺	
Sodium laureth sulphate (may contain carcinogens 1,4-dioxane and ethylene oxide)	Surfactant Detergent	☹	
Sodium lauroyl sarcosinate (may be of ANIMAL origin)	Surfactant Antistatic agent	☹?	

Potential Effects	Cosmetic Uses	Other Uses
Considered safe in cosmetic use; mild skin and eye irritation	Bar soap, body wash, skin scrubs	
See Fluoride	Cosmetics, toothpastes, dentifrices	Cigarettes
Dermatitis of the scalp; ingestion can cause vomiting, hypotension, diarrhoea and collapse; may be hazardous to the environment, especially aquatic organisms	Shampoo, soap, hair straightener, liquid face powder	Pesticides
May release formaldehyde; NIH could not locate any studies for safety	Cosmetics	
Mild reactions in sensitive people	Shampoo, hair conditioner	
Mild eye and skin irritation; can cause the formation of nitrosamines; toxic to aquatic organisms; see Nitrosamines	Shampoo, toothpaste, bath gel, bubble bath, liquid hand and body wash	Dishwashing liquid
See Sarcosines	Hair conditioner	

Names	Function	Code	
Sodium lauryl sulphate (may be prepared synthetically by sulfation of lauryl alcohol then neutralisation with sodium carbonate)	Surfactant Denaturant Emulsifier	☹	
Sodium lauryl sulphoacetate	Surfactant	☺	
Sodium methyl cocoyl taurate (of ANIMAL origin, ox bile)	Emulsifier Surfactant	☺?	
Sodium myreth sulphate (may be of ANIMAL origin)	Emulsifier	☺	
Sodium myristoyl sarcosinate (may be of ANIMAL origin)	Surfactant Antistatic	☺?	
Sodium C14-C16 olefin sulfonate	Surfactant	☺?	

	Potential Effects	Cosmetic Uses	Other Uses
	Skin, eye and mucous membrane irritation; dry skin; eczema; mouth ulcers; liver and gastrointestinal toxicity; on NIH hazards list; teratogen; toxic to aquatic organisms	Bubble bath, hair conditioner, liquid hand and body wash, shampoo, toothpastes, moisturiser	Cake mix, dried egg products, marshmallows, industrial cleaning products
	Mild to strong skin irritation; slight eye irritation; slightly toxic to rats in oral doses	Cream shampoo, cleansing cream, bath bombs	
	May cause formation of nitrosamines; see Nitrosamines	Cosmetics	
	Mild to moderate eye irritation in animal studies	Shampoo	
	See Sarcosines	Moisturiser	
	May cause skin irritation; hair dryness and denaturing; may cause nitrosamine formation	Cosmetics, hair conditioner, shampoo	

Names	Function	Code	
Sodium oleth sulphate (may be of ANIMAL origin)	Emulsifier	☹	
Sodium silicate (water glass)	Anticaking agent	☺?	
Sodium stearate (may be of ANIMAL origin)	Emulsifier Surfactant	☺	
Sorbitan laurate	Emulsifier	☺	
Sorbitan oleate (may be of ANIMAL origin)	Emulsifier Plasticiser	☺?	
Sorbitan palmitate (may be of ANIMAL origin)	Emulsifier	☺	
Sorbitan stearate (may be of ANIMAL origin)	Emulsifier	☺	

	Potential Effects	Cosmetic Uses	Other Uses
	May be contaminated with ethylene oxide and/or 1,4-dioxane (see both for effects)	Cosmetics	
	Can cause skin and mucous membrane irritation; vomiting and diarrhoea when ingested	Barrier cream, soap, depilatories	Preserving eggs, laundry detergent
	Non-irritating to the skin; safety is under review	Toothpastes, shaving lather, soapless shampoo	
	Generally recognised as safe; may cause contact hives	Cosmetic cream and lotion	Foods
	Generally recognised as safe; may cause contact hives and allergic reactions	Cosmetics, eye make-up	
	Generally recognised as safe; may cause contact dermatitis	Shampoo, hair conditioner, cosmetic cream	
	Generally recognised as safe; may cause contact hives	Shampoo, suntan lotion, deodorant, toothpaste, soap	

Names	Function	Code	
Sorbitol (may be synthetic)	Humectant	☺	
Soybean oil (also known as soyabean oil; may be GM)	Emollient	☹?	
Soytrimonium chloride (likely to be GM)	Preservative Emulsifier	☹?	
Spearmint oil (oil of spearmint)	Fragrance Flavouring	☺	
Spermaceti (of ANIMAL origin, from sperm whales)	Emollient	☺	
Squalane (may be of ANIMAL origin – shark liver oil)	Lubricant Emollient	☺☺	
Squalene (may be of ANIMAL origin – shark liver oil)	Emollient Antistatic	☺☺	

	Potential Effects	Cosmetic Uses	Other Uses
	Considered non-toxic when used on the skin	Shampoo, toothpaste, hand lotion	See Sorbitol (E420) section 1
	Goitre from excess consumption; flatulence; indigestion; allergic reactions; pimples; hair damage in topical use	Soap, shampoo, moisturiser, bath oil	Soy sauce, margarine, soy products
	See Quaternary Ammonium Compounds	Cosmetics	
	Considered to have beneficial health effects; may cause skin rash	Perfume, toothpaste, perfumed cosmetics	Chewing gum, cigarettes
	Considered non-toxic but may become rancid and cause skin irritation	Shampoo, cold cream	
	Generally considered safe in cosmetic use	Skin and hair cosmetics	
	Considered to have beneficial health effects	Skin care products, hair dye, fixative in perfume	Shark liver oil, supplements

Names	Function	Code	
Starch (unmodified; found in many plants; may be GM)	Thickener	☺?	
Starch – Modified (starch treated with sodium hydroxide, propylene oxide, aluminium sulphate and others; may be GM)	Thickener Binder	☺?	
Stannous fluoride (solution of tin in hydrofluoric acid; on Canadian Hotlist)	Oral care agent	☹☹	
Stearalkonium chloride (may be of ANIMAL origin)	Preservative	☹	
Stearamide DEA (may be of ANIMAL origin)	Opacifier Antistatic agent	☹	
Stearamide MEA (may be of ANIMAL origin)	Opacifier Antistatic agent	☹	

	Potential Effects	Cosmetic Uses	Other Uses
	Poorly digested; when used in cosmetic products acne; dermatitis; hay fever	Dusting powder, baby powder, dry shampoo	Processed foods
	Safety concern about body's resistance to chemicals used to modify starch especially babies; diarrhoea in babies; when used in cosmetic products acne; dermatitis; hay fever	Cosmetic products	Processed foods, baby foods, cigarettes
	See Fluoride	Dentifrices	
	Mild skin irritation; severe eye irritation; dermatitis; may contain DEA; see Quaternary Ammonium Compounds and Diethanolamine	Hair conditioner	
	DEA-related ingredient; see Diethanolamine	Shampoo, hair conditioner	
	DEA-related ingredient; see Diethanolamine	Shampoo, hair conditioner	

Names	Function	Code	
Stearamidopropyl betaine (may be of ANIMAL origin)	Antistatic agent	😐?	
Stearamidopropyl dimethylamine (may be of ANIMAL origin)	Emulsifier Antistatic agent	😐?	
Stearic acid (may be of ANIMAL origin)	Emulsifier Anticaking agent	😐?	
Stearoyl sarcosine (may be of ANIMAL origin)	Antistatic agent	😐?	
Stearyl alcohol (of ANIMAL origin)	Emollient Opacifier	😊	
Stevia and stevioside (Brazilian herb and extract; banned in some countries)	Oral care agent	😊	
Styrene (derived from ethylbenzene)	Binder	😟	

	Potential Effects	Cosmetic Uses	Other Uses
	See Quaternary Ammonium Compounds	Hair conditioner	
	Allergic dermatitis; may promote the formation of nitrosamines; see Nitrosamines	Hair conditioner	
	May cause allergic reactions in people with sensitive skin; health effects not adequately investigated	Deodorant, hand cream, barrier cream, soap	Chewing gum base, suppositories
	See Sarcosines	Shaving gel	
	May cause allergic reactions and contact dermatitis in people with sensitive skin	Depilatories, hair rinse, moisturiser, shampoo	Pharmaceuticals
	No adverse effects reported in humans, some reports of adverse reactions in animals	Dental care products	Dietary supplements
	Liver, blood, endocrine, kidney and neurotoxicity; teratogenic; carcinogenic	Manufacture of cosmetic resins	Chewing gum, manufacture of plastics

Names	Function	Code	
Styrene/PVP copolomer (from vinyl pyrrolidone and styrene monomers)	Film former Opacifier	☹	
Sulphites (sodium, potassium and ammonium)	Preservative Antioxidant	☹	
Sulfonamide (sulfanilamide; on Canadian Hotlist)	Antibiotic	☹	
Super oxide dismutase (an antioxidant enzyme found in the body)	Biological additive Antioxidant	☺☺	
Talc (naturally occurring mineral)	Anticaking agent Absorbent	☹	
Talcum powder (may contain boric acid)	Absorbent	☹	
TEA compounds	Various	☹	

	Potential Effects	Cosmetic Uses	Other Uses
	See Styrene and Polyvinylpyrrolidone	Liquid eyeliners	
	Asthma; anaphylactic shock; skin rash; nausea; stomach irritation; diarrhoea; swelling; destroy vitamin B1	Hair waves, hair dye, artificial tanning products,	Various foods and beverages, cellophane for food packaging
	Itching; skin rash; swelling; hives; kidney toxicity; teratogenic; on NIH hazards list	Cosmetics, nail polish	Antibiotic to treat bacterial and fungal infections
	Considered to have beneficial effects when applied to the skin	Hair care products, skin cream	Encapsulated and inject-able pharma-ceuticals
	Lung irritation; pneumonia; cough; vomiting; ovarian and lung cancer; carcinogenic	Face cream, baby powder, eye make-ups	
	See Talc above and (E553b) in section 1	See Talc above and (E553b) in section 1	
	See Triethanolamine	Personal care and cosmetic products	

Names	Function	Code	
TEA lauryl sulphate	Surfactant	🙁	
Terpineol	Flavouring Denaturant Solvent	🙁	
Tertiary butylhydroquinone (on Canadian Hotlist)	Antioxidant	🙁	
Tetrabromofluorescein	Colourant	🙁	
Theobroma oil (derived from cacao bean)	Emollient Botanical	🙂	
Theobromine (alkaloid closely related to caffeine)	Botanical	🙂?	
Thimerosal (mercury; on Canadian Hotlist)	Preservative	🙁🙁	

Potential Effects	Cosmetic Uses	Other Uses
See Triethanolamine and Sodium Lauryl Sulphate	Hair care products, mudpacks	
Pneumonitis; eye and mucous membrane irritation; on NIH hazards list; CNS depression	Perfume, hairspray, soap, aftershave, roll-on deodorant	Bleach, laundry detergent, cigarettes
Allergic reactions; contact dermatitis; birth defects in animals; carcinogenic	Cosmetics; lipstick, eye make-up	Foods
Photosensitivity; inflammation of lips; respiratory and gastrointestinal symptoms	Indelible lipstick, nail polish	Dyeing of wool, silk and paper
Allergic reactions in some people; acne	Soap, cosmetics	Confection-ery, pharma-ceuticals
Stimulates the CNS; atrophy of the testicles; endocrine, liver and neurotoxicity; teratogen	Skin conditioner in cosmetics	Chocolate, blood vessel dilator
Allergic reactions; contact dermatitis; see Mercury Compounds	Eye preparations	Filler in vaccines

Names	Function	Code	
Thiourea (made by heating a derivative of ammonium cyanide; on Canadian Hotlist)	Preservative Additive	☹☹	
Thymol (obtained from essential oil of lavender and others)	Additive Fragrance	☺?	
Titanium dioxide (occurs naturally; may contain nanoparticles)	Colourant Opacifier	☺?	
Toluene (derived from petroleum or by distilling Tolu balsam, a plant extract)	Solvent	☹☹	
Toluene-2, 5-diamine (on Canadian Hotlist)	Colourant	☹☹	
Toluene-3, 4-diamine	Colourant	☹☹	

COSMETIC INGREDIENTS

	Potential Effects	Cosmetic Uses	Other Uses
	Skin irritation; allergic reactions; cardiovascular; immuno- and reproductive toxicity; carcinogenic; on NIH hazards list; toxic to aquatic organisms	Hair dye, hair preparations, cosmetics	Photography, dye, wet suits, silver polish
	Allergic reactions; ingestion can cause nausea, vomiting, dizziness; neurotoxicity	Cosmetics, aftershave, mouthwash	Food flavouring
	See Titanium Dioxide (E171) in section 1; see also Nanoparticles	Sunscreen, bath powder, barrier cream	Colouring on foods, paints, marker ink
	Cardiovascular, respiratory, kidney, liver, developmental, reproductive, neuro- and immunotoxicity; eye and skin irritation; decreased learning ability; brain damage; toxic to aquatic organisms	Hair gel, perfume, nail polish, hair dye, hairspray	Removing odours in cheese, metal cleaner, glue
	Harmful to aquatic organisms; see Toluene	Hair dye	
	Harmful to aquatic organisms; see Toluene	Hair dye	Dye for furs, textiles, leather

Names	Function	Code	
Tretinoin (retinoic acid from vitamin A; may be of ANIMAL origin)	Skin improver	☺?	
Tribromosalan (on Canadian Hotlist)	Antiseptic Fungicide	☹	
Trichloroethane (methyl chloroform)		☹☹	
Triclocarban (prepared from aniline, a benzene derivative)	Preservative	☹	
Triclosan (may contain toxic chemicals; on Canadian Hotlist)	Preservative	☹	
Triethanolamine (TEA)	Buffer Coating additive	☹	

	Potential Effects	Cosmetic Uses	Other Uses
	Considered to have beneficial health effects; may cause skin peeling, chapping, blistering and swelling	Anti-wrinkle cream, anti-ageing cream	Acne treatments
	Prohibited in cosmetics in the USA in 2000 as it may cause photoallergies	Soap, medicated cosmetics	
	Severe mucous membrane and eye irritation; liver, neuro- and cardiovascular toxicity; cardiac arrest; vomiting; teratogen; harmful to aquatic organisms	Cosmetics, nail polish	Correction fluid, degreaser, glue, spot remover, detergent
	Photoallergic reactions; convulsions; prolonged use may cause cancer	Soap, medicated cosmetics, deodorant	
	Allergic reactions; contact dermatitis; toxic by ingestion; liver damage in animals	Antiperspirant, deodorant soap mouthwash	Household products, drugs
	Allergic contact dermatitis; skin irritation; may react with nitrites to form nitrosamines; on NIH hazards list	Hand and body lotion, hair conditioner, 'no rinse' shampoo	Coating on fresh fruit and vegetables, detergents

Names	Function	Code	
Trisodium phosphate (from phosphate rock)	Buffer Chelating agent	😐?	
Urea (found in urine; may be of ANIMAL origin; may be synthetic; on Canadian Hotlist)	Humectant Antistatic agent	😐?	
Vanillin (made from eugenol or waste from the wood pulp industry)	Additive Flavouring	😐?	
Waxes (from petroleum, animals, plants and insects; can contain pesticides; may be of ANIMAL origin)	Film former Emollient	😐?	
Wintergreen oil	Flavouring Denaturant	☹	
Yucca extract (derived from a plant grown in south-western USA)	Foaming agent Botanical	☺☺	

Potential Effects	Cosmetic Uses	Other Uses
Can cause skin irritation; neurotoxicity	Shampoo, bubble bath, cuticle softener	Additive in foods, pesticides
Thinning of the epidermis; allergic reactions in some people; dermatitis; alleviates dry skin	Skin cream and lotion, mouthwash, moisturiser	Browning agent in baked goods like pretzels, cigarettes
Skin irritation; eczema; skin pigmentation; contact dermatitis; on NIH hazards list	Perfume	Chocolate, cheese, candy, cigarettes
Generally safe in cosmetic use; may cause allergic reactions depending on source and purity	Cosmetics, hair-grooming preparations, lipstick, hair straightener	Coatings on fresh fruit and vegetables, packaging materials
Harmful to aquatic organisms; see Methyl Salicylate	See Methyl Salicylate	See Methyl Salicylate
Considered to have beneficial health effects	Shampoo, organic cosmetics	Root beer flavouring

Names	Function	Code	
Zinc chloride (soluble zinc salt)	Oral care agent	☹	
Zinc myristate (zinc salt of myristic acid)	Opacifier Viscosity control	😐?	
Zinc oxide (may contain nanoparticles)	Opacifier Additive Colourant	😐?	
Zinc stearate (may be of ANIMAL origin)	Colourant	😐?	
Zinc sulphate (from reaction of zinc and sulphuric acid)	Anti-microbial	😐?	
Zirconium (banned in aerosol cosmetic products in the USA; on Canadian Hotlist)	Solvent Abrasive	😐?	

Potential Effects	Cosmetic Uses	Other Uses
Toxic; mild skin irritation; contact dermatitis; can be absorbed through the skin; teratogenic; very toxic to aquatic organisms	Mouthwash, dentifrices	Pesticides
Toxic; may promote acne; nausea and vomiting if ingested	Make-up, nail polish	
Helps protect against UV radiation; may be unsuitable for dry skin; respiratory toxicity; may cause skin eruptions; teratogenic; see Nanoparticles	Baby powder, antiperspirant, shaving cream, calamine lotion, sunscreen, hair products	Used medically as an antiseptic, astringent and protective in skin diseases
Skin and eye irritation; lung problems and pneumonitis	Baby powder, hand cream, face powder	Tablet manufacture
Skin and mucous membrane irritation; allergic reactions; cardiovascular toxicity	Skin tonics, eye lotion, aftershave, shaving cream	Paperboard products
Considered safe in non-aerosol products; toxic by inhalation; respiratory toxicity; contact allergic reactions	Cosmetic cream, antiperspirant, deodorant	Preparation of dye

Genetic Modification in a Nutshell

Genetic Modification (GM), otherwise known as Genetic Engineering (GE), involves taking genes from one species and inserting them into another in an attempt to transfer a desired trait or characteristic. Genes (comprised of sequences of amino acids) are the biological units of heredity, the individual messages that go together to form DNA strands, the blueprints for the thousands of proteins that combine to form the building blocks of all life from bacteria to humans.

Think of it like the book you are reading now. The individual letters (amino acids) are arranged to form words (genes) that produce sentences (proteins), and the sentences link together to eventually become the book (life form).

An example of genetic modification involves taking the genes from say, an arctic fish, which has *antifreeze* properties, and inserting them into a tomato to provide resistance from frost. The unnatural process of genetic modification can lead to unpredictable effects, as it is impossible to guide the insertion of the new genes. Genes do not work in isolation, but in highly complex relationships that are a long way off being fully understood. Any change to the DNA at any point will affect it in ways scientists cannot predict. Traditional breeding techniques operate within established natural boundaries allowing reproduction to take place only between closely related forms. Tomatoes can cross-pollinate with other tomatoes, but not soybeans and definitely not arctic fish; pigs can mate only with pigs and not sheep. The genes in their natural groupings have been finely tuned to work harmoniously together by millions of years of evolution.

Crossing genes between unrelated species that would never crossbreed in nature can give rise to **potential health risks**. One problem with GM foods is their unpredictability. A person may prove unexpectedly allergic to a food he or she has previously eaten safely. In one case, soybeans engineered with a gene from Brazil nuts caused allergic reactions in people sensitive to the nuts. People who are hyper-allergenic or environmentally sensitive would be well advised to avoid GM foods. Most genes being introduced into GM plants have never been part of the food supply so we can't know if they are likely to be allergenic. In 1989 there was an outbreak of a new disease in the USA contracted by over 5,000 people and traced back to a batch of L-tryptophan food supplement produced with GM bacteria. Even though it contained less than 0.1 per cent of a highly toxic compound, 37 people died and 1,500 were left with permanent disabilities. More may have died, but the American Center for Disease Control stopped counting in 1991.

Health-risk assessment of GM foods compares only GM and non-GM equivalent varieties of a few known components (e.g. nutrients, toxins and allergens). If things match up then all is assumed to be well. Short-term animal feeding trials are conducted in some cases. **All research is done by the biotech companies themselves, or companies they employ**. Then government approval committees judge whether they believe that the documented evidence of safety is convincing. No evidence from human trials for either toxicity or allergy testing is required. No independent checks of the company's claims are required and test results are rarely published for scientific review.

Professor Joe Cummins, professor emeritus of genetics at the University of Western Ontario, believes there is a cynical agenda behind the lack of proper testing. 'The failure to test may provide some protection in the courts against lawsuits by those maimed or crippled by the foods. Most ill effects from food and allergies are not easily quantified until after the disaster. At best, there may be a small but marked increase in autoimmune disease and allergy associated with the (GM) foods. At worst, major outbreaks of illness could be observed and will be difficult to trace to the unlabelled foods.' he said.

The multinational companies that create GMOs claim that their GM technology will feed the starving populations of the world, provide better yields for producers and supply us with healthier and more nutritious foods. Scientists, agriculturalists and medical experts around the globe have openly refuted each of these claims. On the other hand, research shows that sustainable agriculture results in higher productivity and yields, especially in the Developing World. Continued practice results in better quality soils, a reduction in soil erosion, a cleaner and safer environment and a reduction in pesticide use without a subsequent increase in pests. Sustainable agriculture leads to healthier and tastier foods with higher nutritional values. Anyone who has eaten home-grown strawberries grown without the use of artificial chemicals will attest to this.

Patents give a huge incentive to the biotechnology industry to create new GM organisms. Patents mean money, lots of it, and most patents last for 17-20 years. There are currently patents approved or pending for at least 200 GM animals, including fish, cows, mice and pigs. There are also patents on varieties of seeds and plants, as well as unusual genes and cell lines from indigenous peoples.

Genetically modified foods have been steadily and insidiously invading our food supply since the 1980s. GM soybean is in more than 60 per cent of all processed foods as vegetable oil, soy flour, lecithin and soy protein. GM maize is in about 50 per cent of processed foods as corn, cornstarch, corn flour and corn syrup. GM tomato puree is sold in some supermarkets and GM enzymes are used throughout the food processing industry. Government regulations on labelling exclude at least 95 per cent of the products containing GM ingredients because they ignore derivatives.

Currently, **the best way to avoid products containing GM additives and ingredients** is to purchase only products labelled 'GM-free' or 'Not Genetically Modified'. Certified Organic and Certified Biodynamic products are also GM-free. However, even with the best intentions, companies attempting to exclude GM ingredients from their products have found contamination from GMO's as a result of cross-pollination. If we don't stop GM crops, organic produce will be permanently contaminated with no way back.

If you are opposed to genetically modified products, there are several ways that you can be pro-active. Buy only products labelled Certified Organic, GM-free etc; ask your local supermarket to stock only non-GM products; contact the manufacturers of food products you buy and if they do use genetically modified ingredients, request that they go GM-free.

Once released, genetically modified organisms become part of our ecosystem, unlike some other forms of pollution which can be contained or which may decrease over time. Any mistakes we make now will be passed on to all future generations of life. With governments capitulating to commercial interests, **it is up to us to act**.

APPENDIX

Safe and/or Beneficial Ingredients Used in Cosmetics and Personal Care Products

Listed below is a small selection of some of the safe and/or beneficial ingredients to look for when shopping for personal care products and cosmetics. It is wise to select products with organic (preferably certified organic) ingredients and avoid those with too many synthetic chemicals, especially ones with this ☹ or this ☹☹ symbol. If the product is not certified organic, check the label for evidence that the ingredients are GM-free.

Name	Function
Aloe Vera	botanical
Avocado Oil	natural emollient
Ascorbic Acid (vitamin C)	natural preservative
Candelilla Wax	natural emulsifier
D-Alpha Tocopherol Acetate (vitamin E)	natural preservative
Evening Primrose Oil	botanical
Grapefruit Seed Extract	natural preservative
Hemp Oil	botanical
Honeysuckle Extract	natural fragrance
Jojoba Butter	natural emollient

Jojoba Oil	natural emulsifier
Lecithin (GM-free vegetable origin)	natural humectant
Macadamia Oil	natural emollient
Olive Oil (Castile) Soap	natural surfactant
Panthenol, Dexpanthenol (vitamin B5)	natural emollient
Purified Water	natural solvent
Quince Seed	natural emulsifier
Rice Bran	natural emulsifier
Rosehip Seed Oil	natural emollient
Rosemary Extract	natural preservative
Rose Water	natural perfume
Saffron	natural colouring
Sclerotium Gum	natural emulsifier
Shea Butter	natural emollient
Soapwort	natural surfactant
Stevia	natural sweetener
Vitamin A Palmitate	natural preservative
Xanthan Gum (GM-free)	natural emulsifier
Yucca Extract	natural surfactant

Glossary

ABRASIVE: A substance added to cosmetic products either to remove materials from various body surfaces or to aid mechanical tooth cleaning and improve gloss.

ABSORBENT: A substance added to cosmetic products to take up water and/or oil-soluble dissolved or finely dispersed substances.

ACETYLATED: An organic compound that has had its water removed by heating with acetic anhydride or acetyl chloride. Both these chemicals are hazardous.

ADDITIVE: A substance added to cosmetic products, often in relatively small amounts, to impart or improve desirable properties or minimise undesirable properties.

ALLERGEN: Any substance capable of provoking an inappropriate immune response in susceptible people, but not normally in others.

ALLERGIC CONTACT DERMATITIS: A skin rash caused by direct contact with a substance to which the skin is sensitive. Symptoms may occur anywhere from seven days to many years after repeated low-level exposures, as occurs with cosmetics and personal care products.

ALLERGIC REACTION: An adverse immune system response involving unusual sensitivity to the action of various environmental stimuli. These stimuli do not normally cause symptoms in the majority of the population.

AMINES: A class of organic compounds derived from ammonia.

ANTICAKING AGENT: A substance used in granular foods like salt or flour to assist free flowing.

ANTICORROSIVE: Chemicals added to cosmetics to prevent corrosion of the packaging or the machinery used in the manufacture of the cosmetic.

ANTIFOAMING AGENT: A substance added to foods or cosmetics to prevent excessive frothing or foaming, reduce the formation of scum or prevent boiling over during manufacture.

ANTIMICROBIAL: A substance added to a cosmetic product to help reduce the activities of micro-organisms on the skin or body.

ANTIOXIDANT: A substance added to foods or cosmetics to prevent changes or spoiling due to exposure to air. May be natural or synthetic.

ANTISTATIC: A substance used to reduce static electricity by neutralising electrical charge on a surface.

AZO DYE: A very large class of dye made from diazonium compounds and phenol. Many azo dyes are thought to be carcinogenic when used in foods.

BINDER: A substance added to a solid cosmetic mixture to provide cohesion.

BIOLOGICAL ADDITIVE: A substance, derived from a biological origin, added to a cosmetic product to achieve a specific formulation feature.

BLEACHING AGENT: A substance used to artificially bleach and whiten flour. A substance used in a cosmetic product to lighten the shade of hair or skin.

BOTANICAL: A substance, derived from plants, added to a cosmetic product to achieve a specific formulation feature.

BUFFER: A substance added to a food or cosmetic product to adjust, maintain or stabilise the acid/alkali (pH) balance.

CANADIAN HOTLIST: Information about cosmetic ingredients that have the potential for adverse effects or which have been restricted or banned.

CARCINOGEN: A cancer-causing substance. IARC and NTP list carcinogens in 3 categories. 1 = confirmed human carcinogen; 2 = probable human carcinogen; 3 = possible human carcinogen.

CARCINOGENIC: A substance that is capable of causing cancer.

CARDIOVASCULAR/BLOOD TOXICITY: Adverse effects on the cardio-vascular or haematopoietic systems that result from exposure to chemical substances. Exposure can contribute to a variety of diseases; including elevated blood pressure (hypertension), hardening of the arteries (arteriosclerosis), abnormal heartbeat (cardiac arrhythmia) and decreased blood flow to the heart (coronary ischaemia). Exposure can also reduce the oxygen carrying capacity of red blood cells, disrupt important immunological processes carried out by white blood cells and induce cancer.

CHELATING AGENT: A substance added to a food or cosmetic product to react and form complexes with metal ions that could affect stability and/or appearance.

CIR EXPERT PANEL: A body set up in 1976 by the Cosmetic, Toiletry and Fragrance Association to review the safety of ingredients used in cosmetics.

CLARIFYING AGENT: A substance that removes small amounts of suspended particles from liquids.

CNS: Central Nervous System – our body's major communication network.

COAL TAR DYE: Dye that were once made from coal tar but are now commercially produced by a synthetic process. These dye are extremely complex chemical compounds, which have had inadequate testing and often contain toxic impurities.

CONTACT DERMATITIS: *See allergic contact dermatitis.*

COSMETIC ACNE: Acne caused by applying cosmetics to the skin.

CYTOTOXIN: A substance that is poisonous to cells.

DENATURANT: A poisonous or unpleasant substance added to alcoholic cosmetics to prevent them being ingested. It is also a substance that changes the natural qualities or characteristics of other substances.

DENTIFRICES: Pastes, powders or liquids for cleaning the teeth.

DEPILATORY: A substance or agent used to remove unwanted body hair.

DERMATITIS: Inflammation of the skin with pain, redness, burning or itching and fluid build-up.

DEVELOPMENTAL TOXICITY: Adverse effects on the developing foetus that result from exposure to chemical substances. Developmental toxicants, sometimes called teratogens, include agents that induce structural malformations and other birth defects, low birth weight, metabolic or biological dysfunction and psychological or behavioural deficits that become manifest as the child grows.

DILUENT: A substance used to dilute or dissolve other additives.

DPIM: 'Dangerous Properties of Industrial Materials.' Ed. Sax & Lewis.

ECZEMA: Wet or dry inflammation of the skin causing redness, pain, itching, scaling, peeling, blistering etc.

EDF: Environmental Defense; provides information on chemicals.

EMOLLIENT: A substance used to soften and soothe the skin.

EMULSIFIER: A substance used in food or cosmetic products to stabilise mixtures and ensure consistency.

EMULSION STABILISER: A substance added to a cosmetic product to help the process of emulsification and to improve formulation stability and shelf life.

ENDOCRINE TOXICITY: Adverse effects on the structure and/or functioning of the endocrine system that result from exposure to chemical substances. The endocrine system is composed of many organs and glands that secrete hormones directly into the bloodstream including the pituitary, hypothalamus, thyroid, adrenals, pancreas, thymus, ovaries and testes. Compounds that are toxic to the endocrine system may cause diseases such as hypothyroidism, diabetes mellitus, hypoglycaemia, reproductive disorders and cancer.

GLOSSARY

EPA: Environmental Protection Agency

ESTER: A compound formed when an acid reacts with an alcohol by the elimination of water.

ETHOXYLATION: The addition of ethyl (from the gas ethane) and oxygen to a degreasing agent to make it less abrasive and cause it to foam more.

FDA: Food and Drug Administration (USA). It is part of the Public Health Service of the US Department of Health and Human Services.

FILM FORMER: A substance added to a cosmetic product to produce, when applied, a continuous film on skin hair or nails.

FLAVOUR ENHANCER: Chemicals that enhance the taste or odour of food without contributing any taste or odour of their own.

FLAVOURING: The largest category of food additives. Over two thousand synthetic and natural flavourings added to foods to impart the desired flavour.

FRAGRANCE: Any natural or synthetic substance used to impart an odour to a product.

FUNGICIDE: A substance used to kill or inhibit the growth of fungi.

GELLING AGENT: A substance that is capable of forming a jelly.

GM: Abbreviation for Genetically Modified.

GMO: Genetically Modified Organism.

GLAZING AGENT: A substance used to provide a shiny appearance or a protective coat to a food.

GRAS: Generally Recognised As Safe. A list, established by American Congress in 1958, of substances added to food over a long time.

HAZARDOUS CHEMICAL AGENTS: 1.Those chemical agents known to have undesirable biological effects, either acutely or chronically, reasonable regard being given to the size of the dose, duration and type of exposure and the physical state of the compound required to produce such effects. 2.Those agents for which toxicity information is not available but are highly suspect for reasons of similarity in chemical structure or function to known toxic agents. 3. Those agents that are explosive or violently reactive.

HERBICIDE: A substance used to kill or inhibit the growth of unwanted plants.

HIVES: An allergic disorder marked by raised, fluid-filled patches of skin or mucous membrane, usually accompanied by intense itching. Also known as nettle rash and urticaria.

HUMECTANT: A substance used to hold and retain moisture to prevent a food or product from drying out.

HYDROGENATED: Liquid oils in food and cosmetic products are converted to semisolid fats at room temperature by adding hydrogen under high pressure. Hydrogenated fats and oils contribute to cancer, heart disease and atheroma.

HYDROLYSED: Turned partly into water as a result of a chemical process.

IMMUNOTOXICITY: Adverse effects on the functioning of the immune system that result from exposure to chemical substances. Altered immune function may lead to the increased incidence or severity of infectious diseases or cancer, since the immune systems ability to respond adequately to invading agents is suppressed. Toxic agents can also cause autoimmune diseases, in which healthy tissue is attacked by an immune system that fails to differentiate self-antigens from foreign antigens.

INTERMEDIATE: A chemical substance found as part of a necessary step between one organic compound and another.

KIDNEY TOXICITY: Adverse effects on the kidney, ureter or bladder that result from exposure to chemical substances. Some toxic agents cause acute injury to the kidney, while others produce chronic changes that can lead to end-stage renal failure or cancer. The consequences of renal failure can be profound, sometimes resulting in permanent damage that requires dialysis or kidney transplantation.

LIVER/GASTROINTESTINAL TOXICITY: Adverse effects on the structure and/or functioning of the gastrointestinal tract, liver, or gall bladder that result from exposure to chemical substances. The liver is frequently subject to injury induced by chemicals, called hepatotoxins, because of its role as the body's principal site of metabolism.

MATERIAL SAFETY DATA SHEETS (MSDS): Data compiled by manufacturers of chemicals providing information on health hazards and safe handling procedures.

MILIARIA: Acute itchy skin condition occurring as an eruption of spots or blisters resembling millet seeds.

MODIFIER: A substance that induces or stabilises certain shades in hair colouring.

MUSCULOSKELETAL TOXICITY: Adverse effects on the structure and/or functioning of the muscles, bones and joints that result from exposure to chemical substances. Exposure to toxic substances such as coal dust and cadmium has been shown to cause adverse changes to the musculoskeletal system. The bone disorders arthritis, fluorosis and osteomalacia are among the musculoskeletal diseases that can be induced by occupational or environmental toxicants.

MUTAGEN: Any substance that induces mutation or permanent changes to genetic material (DNA) of cells.

MUTAGENIC: Capable of causing mutations. Can be induced by stimuli such as certain food chemicals, pesticides and radiation.

NANOPARTICLES: Anything smaller than 100 nanometres (a nanometre is a billionth of a metre) in size or more than 800 times smaller than a human hair. They can enter the bloodstream and cross the blood-brain barrier.

NECROSIS: Cell death.

NEUROTOXICITY: Adverse effects on the structure or

functioning of the central and/or peripheral nervous system that result from exposure to chemical substances. Symptoms of neurotoxicity include muscle weakness, loss of sensation and motor control, tremors, alterations in cognition and impaired functioning of the autonomic nervous system.

NIH: National Institutes of Health. Provides a data bank of hazardous chemicals.

NIOSH: The National Institute of Occupational Safety and Health, which is the research arm of the US Occupational Safety and Health Administration (OSHA).

NITROSAMINES: Potential carcinogenic compounds formed when an amine reacts with a nitrosating agent or substances containing nitrites.

NITROSATING AGENT: A substance capable of introducing nitrogen and oxygen molecules into a compound that may cause the compound to form potential carcinogenic nitrosamines.

NRC: Not recommended for children.

NTP: National Toxicology Program (USA). Information on chemical toxicity.

OPACIFIER: A substance added to a shampoo or other transparent or translucent liquid cosmetic product to make it impervious to visible light or nearby radiation.

ORAL CARE AGENT: A substance added to a personal care product for the care of the oral cavity.

OXIDISING AGENT: A substance added to a food or cosmetic product to change the chemical nature of another substance by adding oxygen.

PHOTOALLERGY: *See photosensitivity*.

PHOTOSENSITIVITY: A condition in which the application to the body or ingestion of certain chemicals causes skin problems (rash, pigmentation changes, swelling etc) when the skin is exposed to sunlight. Also know as photoallergy.

PHOTOTOXICITY: Reaction to sunlight or ultraviolet light resulting in inflammation.

PLASTICISER: A substance added to impart flexibility and workability without changing the nature of a material.

PRESERVATIVE: A substance added to food and cosmetic products to inhibit the growth of bacteria, fungi and viruses.

PROPELLANT: A gas used to expel the contents of containers in the form of aerosols.

REAGENT: A substance used for the detection of another substance by chemical or microscopic means.

REDUCING AGENT: A substance added to food and cosmetic products to decrease, deoxidise or concentrate the volume of another substance.

REPRODUCTIVE TOXICITY: Adverse effects on the male and female reproductive systems that result from exposure to chemical substances. Reproductive toxicity may be expressed as alterations in sexual behaviour, decreases in fertility or loss of the foetus during pregnancy. A reproductive toxicant may interfere with the sexual functioning or reproductive ability of exposed individuals from puberty throughout adulthood.

RESPIRATORY TOXICITY: Adverse effects on the structure or functioning of the respiratory system that result from exposure to chemical substances. The respiratory system consists of the nasal passages, pharynx, trachea, bronchi and lungs. Respiratory toxicants can produce a variety of acute and chronic pulmonary conditions, including local irritation, bronchitis, pulmonary oedema, emphysema and cancer.

RTECS: The Registry of Toxic Effects of Chemical Substances.

SENSITISATION: Heightened immune response following repeated contact with an allergen.

SEQUESTRANT: A substance capable of attaching itself to unwanted trace metals such as cadmium, iron and copper that cause deterioration in food and cosmetic products by advancing the oxidation process.

SOLVENT: A substance added to food and cosmetic products to dissolve or disperse other components.

STABILISER: A substance added to a product to give it body and to maintain a desired texture.

SURFACE ACTIVE AGENT: A substance that reduces surface tension when dissolved in solution. These agents fall into three categories: detergents, wetting agents and emulsifiers.

SURFACTANT: A wetting agent that lowers the surface tension of a liquid substance allowing it to spread out and penetrate more easily. Surfactants fall into four main categories – anionic, non-ionic, cationic and amphoteric.

TENDERISER: A substance or process used to alter the structure of meat to make it less tough and more palatable.

TERATOGEN: *See developmental toxicity.*

TERATOGENIC: Capable of causing defects in a developing foetus.

TEXTURISER: A substance used to improve the texture of various foods or cosmetics.

THICKENER: A substance used to add viscosity and body to foods, lotion and cream.

UV ABSORBER: A substance added to a cosmetic product to filter ultra-violet (UV) rays so as to protect the skin or the product from the harmful effects of these rays.

VISCOSITY CONTROLLING AGENT: A substance added to a cosmetic product to increase or decrease the viscosity (flowability) of the finished product.

XENOESTROGEN: An environmental compound that has oestrogen-like activity thereby mimicking the properties of the hormone oestrogen.

Bibliography

Agency for Toxic Substances and Disease Registry, (ATSDR)

American Academy of Dermatology

Antczak, Dr Stephen & Gina, 'Cosmetics Unmasked', *Thorsons,* 2001

Australian Consumers Association

Cancer Prevention Coalition

Center for Science in the Public Interest (CSPI)

Commonwealth Scientific and Industrial Research Organisation (CSIRO), Australia

Crumpler, Diane, 'Chemical Crisis', *Scribe Publications*

Day, Phillip, 'Cancer – Why We're Still Dying to Know the Truth', *Credence Publications,* 2000

Day, Phillip, 'Health Wars', *Credence Publications,* 2001

Department of Food Science and Technology (UK)

Dingle, Peter and Toni Brown, 'Dangerous Beauty – Cosmetics and Personal Care' *Healthy Home Solution,* 1999

Epstein, Samuel S. M.D., 'Unreasonable Risk' *Environmental Toxicology,* 2002

Environmental Defense

Environmental Protection Agency, (EPA), (USA)

Environmental Working Group

Food and Drug Administration (FDA), (USA)

Food Standards Agency (UK)

Food Standards Australia New Zealand, (FSANZ)

Cummins, Ronnie and Ben Lilliston, 'Genetically Engineered Food – A Self Defence Guide for Consumers' second edition, *Marlow & Company,* 2004

Hampton, Aubrey, 'What's in Your Cosmetics' *Organica Press*

Hanssen Maurice, with Jill Marsden, 'The New Additive Code Breaker' *Lothian*, 1991

In-Tele-Health, Hyperhealth Natural Health & Nutrition CD-ROM 2005 Ed.

International Agency for Research on Cancer, (IARC)

Joint Expert Committee on Food Additives, (JECFA)

Journal of the American Medical Association

Journal of the American College of Toxicology

Lancet, The

Leading Edge Research

Material Safety Data Sheets, (MSDS) from numerous sources

National Center for Environmental Health

National Food Safety Database

National Institutes of Health (NIH), (USA)

National Institute of Occupational Safety and Health (NIOSH)

National Libraries of Medicine (USA)

National Resources Defense Council

National Toxicology Program (USA), (NTC)

Organic Consumers Association

Organic Federation of Australia

Registry of Toxic Effects of Chemical Substances, The, (RTECS)

BIBLIOGRAPHY

Sax & Lewis 'Dangerous Properties of Industrial Materials' Seventh Edition

Sargeant, Doris and Karen Evans, 'Hard to Swallow – The Truth About Food Additives', *Alive Books*, 1999

Steinman, David and Samuel S Epstein 'The Safe Shopper's Bible' *Macmillan*, 1995

Taubert, P.M., 'Silent Killers' *CompSafe Consultancy,* 2001

Taubert, P.M., 'Your Health and Food Additives – 2000 Edition' *CompSafe Consultancy*

Taubert, P.M., 'Read the Label Know the Risks',*CompSafe Consultancy,* 2004

Total Environment Centre, 'A-Z of Chemicals in the Home', 4th edition

Winter, Ruth M.S., 'A Consumer's Dictionary of Cosmetic Ingredients – Sixth Edition' *Three Rivers Press*, 2005

Winter, Ruth M.S., 'A Consumer's Dictionary of Food Additives – Sixth Edition' *Three Rivers Press*, 2004

Useful Internet Resources

The following links to Internet websites have been included here to give you a starting place for doing your own research. All links were accessible at the time of writing.

Organisation	Website Link
Alzheimer's Disease International	www.alz.co.uk
Anaphylaxis Campaign	www.anaphylaxis.org.uk
Asthma UK	www.asthma.org.uk
BDF Newlife (Birth Defects Foundation)	www.bdfnewlife.co.uk
Cancer Prevention Coalition	www.preventcancer.com
Center for Science in the Public Interest	www.cspinet.org
Environmental Defense Fund	www.edf.org
Environmental Working Group	www.ewg.org
Food Standards Agency	www.food.gov.uk

USEFUL INTERNET RESOURCES

Hyperactivity Children's
Support Group   www.hacsg.org.uk

International Agency for
Research on Cancer (IARC)   www.iarc.fr

Material Safety Data Sheets   www.msdssearch.com

Mindfully.Org   www.mindfully.org

National Toxicology Program http://ntp.niehs.nih.gov

Organic Consumers
Association   www.organicconsumers.org

Organic Natural Health   www.health-report.co.uk

PAN Pesticides Database   www.pesticideinfo.org

Women's Environmental
Network   www.wen.org.uk

Visit our website www.thechemicalmaze.com for information on safer foods, cosmetics and personal care products.

www.summersdale.com